Lecture Notes in Mathematics

Edited by A. Dold and B. Eckmann

Subseries: Fondazione C.I.M.E., Firenze
Adviser: Roberto Conti

1337

E. Sernesi (Ed.)

Theory of Moduli

Lectures given at the 3rd 1985 Session of the Centro Internazionale Matematico Estivo (C.I.M.E.) held at Montecatini Terme, Italy, June 21–29, 1985

Springer-Verlag
Berlin Heidelberg New York London Paris Tokyo

Editor

Edoardo Sernesi
Dipartimento di Matematica, Istituto "Guido Castelnuovo"
Università degli Studi di Roma, "La Sapienza"
Piazzale Aldo Moro, 5, 00185 Roma, Italy

Mathematics Subject Classification (1980): 14 H 10; 14 H 15; 14 H 40; 14 J 15; 14 K 25; 57 N 05; 57 R 19

ISBN 3-540-50080-4 Springer-Verlag Berlin Heidelberg New York
ISBN 0-387-50080-4 Springer-Verlag New York Berlin Heidelberg

© Springer-Verlag Berlin Heidelberg 1988
Printed in Germany

Printing and binding: Druckhaus Beltz, Hemsbach/Bergstr.
2146/3140-543210

INTRODUCTION

This volume contains the texts of the three main lecture series given at the CIME session on "Theory of moduli" held in Montecatini during the period 21-29 June, 1985.

The lectures survey some important areas of current research in topology, complex analysis, algebraic geometry, which have as their common denominator the study of moduli spaces. Hopefully, this volume will be a useful reference text on the subject.

Other, more specialized, lectures were also given during the session but they are not reproduced here.

I am very grateful to the three authors, to the other lecturers and to all the participants to the conference for their interest and collaboration.

My thanks go also to the CIME for making the conference possible.

Edoardo Sernesi

TABLE OF CONTENTS

C.I.M.E. Session on "Theory of Moduli"

List of Participants

A. ALZATI, Via G. Tavecchia 47, 20017 Rho, Milano, Italy

E. BALLICO, Scuola Normale Superiore, Piazza dei Cavalieri 7, 56100 Pisa, Italy

L. BRAMBILA PAZ, Depto. de Matematicas Edif. T., Av. Michoacan y la Purisima, Iztapalapa Apdo. Postal 55-534, 09340 México, D.F., México

M. CANDILERA, Ecole Polytechnique, Centre de Mathématiques, 91128 Palaiseau, France

G. CANUTO, Dipartimento di Matematica, Università, Strada Nuova 65, 27100 Pavia, Italy

F. CATANESE, Dipartimento di Matematica, Università, Via Buonarroti 2, 56100 Pisa, Italy

A. COLLINO, Corso Re Umberto 166, 12039 Verzuolo, Cuneo, Italy

P. CRAGNOLINI, Dipartimento di Matematica, Università, Via Buonarroti 2, 56100 Pisa, Italy

V. CRISTANTE, Istituto di Algebra e Geometria, Università, Via Belzoni 7, 35131 Padova, Italy

R. DONAGI, Northeastern University, Department of Mathematics, Huntington Ave., Boston, MA 02115, USA

G. ELENCWAJG, 28 rue Paul Bounin, 06100 Nice, France

G. FERRARESE, Dipartimento di Matematica, Università, Via Carlo Alberto 10,10123 Torino, Italy

F. GAETA, Facultad de Matematica, Universidad Complutense de Madrid, Madrid 3, Spain

G. GONZALEZ-DIEZ, Department of Mathematics, King's College London, Strand, London WCZR 2LS, England

J. HARER, Department of Mathematics, University of Maryland, College Park, Md 27742, USA

H. HELLING, Universität Bielefeld, Fakultät für Mathematik, Postfach 8640 4800 Bielefeld

C.F. HERMANN, Lehrstuhl II für Mathematik, Universität Mannheim, 6800 Mannheim, BRD

S. KILAMBI, Department of Mathematics, University of Montréal, Montréal (Que) M3C-3J7, Canada

A. LANTERI, Dipartimento di Matematica, Università, Via C. Saldini 50, 20133 Milano, Italy

H. MAEDA, Lehrstuhl für Mathematik VI, Universität Mannheim,
6800 Mannheim 1, BRD

L. MAZZI, Corso G. Ferraris 110, 10129 Torino, Italy

A. NANNICINI, Via Cesare Battisti 16, 50068 Montebonello, Firenze, Italy

T. ODA, Max-Planck-Institut für Mathematik, Gottfried-Claren-Str. 26,
5300 Bonn 3, BRD

G. OTTAVIANI, Istituto Matematico, Università, Viale Morgagni 67/A,
50134 Firenze, Italy

A. PAPADOPOULOS, c/o Hautefeuille, 3 rue Jean Moulin, 24800 Thiviers, France

R. PARDINI, Via Verdi 27, 55043 Lido di Camaiore, Lucca, Italy

G. PARIGI, Via Toscanini 50, 50019 Sesto Fiorentino, Firenze, Italy

L. PICCO BOTTA, Via S. Francesco d'Assisi 14, 10093 Collegno, Torino, Italy

G.P. PIROLA, Dipartimento di Matematica, Università, Strada Nuova 65,
27100 Pavia, Italy

K. RANESTAD, Institute of Mathematics, P.O.Box 1053, Blindern, Oslo 3, Norway

F. RECILLAS-JUAREZ, Instituto de Matematicas, Area de la Investigacion Cientifica,
Circuito Exterior, Ciudad Universitaria, México 20, D.F., México

N. RODINO', Via di Vacciano 87, 50015 Grassina, Firenze, Italy

M. SALVETTI, Via G. Leopardi 25, 56010 Ghezzano, Pisa, Italy

W.K. SEILER, Lehrstuhl VI für Mathematik, Universität Mannheim,
6800 Mannheim 1, BRD

M. SEPPALA, University of Helsinki, Department of Mathematics, Hallituskatu 15,
SF-00100 Helsinki, Finland

E. SERNESI, Dipartimento di Matematica, Università, Piazzale A. Moro 5,
00185 Roma, Italy

M. TEIXIDOR I BIGAS, Facultat de Matematique, Universitat de Barcelona,
Gran Via 585, 08007 Barcelona, Spain

C. TRAVERSO, Dipartimento di Matematica, Università, Via Buonarroti 2,
56100 Pisa, Italy

L. VERDI, Istituto Matematico, Università, Viale Morgagni 67/A,
50134 Firenze, Italy

G.G. WEILL, Department of Mathematics, Polytechnic Institute of New York,
333 Jay Street, Brooklyn, New York 112201, USA

MODULI OF ALGEBRAIC SURFACES

F. Catanese* - Università di Pisa**

Contents of the Paper

Introduction

This paper reproduces with few changes the lectures I actually delivered at the C.I.M.E. Session in Montecatini, with the exception of most part of one lecture where I talked at length about the geography of surfaces of general type: the reason for not including this material is that it is rather broadly covered in some survey papers which will be published shortly ([Pe], [Ca 3], [Ca 2]).

Concerning my original (too ambitious) intentions, conceived when I accepted Eduardo Sernesi's kind invitation to lecture about moduli of surfaces, one may notice some changes from the preliminary program: the topics "Existence of moduli spaces for algebraic varieties" and "Moduli via periods" were not treated. The first because of its broadness and complexity (I realized it might require a course on its own, while I mainly wanted to arrive to talk about surfaces of general type), the second too because of its vastity and also for fear of overlapping with the course by Donagi (which eventually did not treat period maps and variation of

---

* A member of G.N.S.A.G.A. of C.N.R., and in the M.P.I. Research Project in Algebraic Geometry.

** The final version of the paper was completed during a visit of the author to the University of California, San Diego.

Hodge structures). Anyhow the first topic is exhaustively treated in Popp's lecture notes ([Po]) and in the appendices to the second edition of Mumford's book on Geometric Invariant Theory ([Mu 2]), whereas the nicest applications of the theory of variation of Hodge structures to moduli of surfaces are amply covered in the book by Barth-Peters-Van de Ven ([B-P-V]).

Also, I mainly treated moduli of surfaces of general type, and fortunately Seiler lectured on the results of his thesis ([Sei 1,2,3]) about the moduli of (polarized) elliptic surfaces: I hope his lecture notes are appearing in this volume.

Instead, the part on Kodaira-Spencer's theory of deformations and its connections with the classical theory of continuous systems started to gain a dominant role after I gave a series of lectures at the Institute for Scientific Interchange (I.S.I.) in Torino on this subject. In fact, after Zappa (cf. [Zp], [Mu 3]) discovered the first example of obstructed deformations, a smooth curve in an algebraic surface, it was hard to justify most of the classical statements about moduli (and in fact, cf. lecture four, some classical problems about completeness of the characteristic system have a negative answer).

Interest in moduli was revived only through the pioneering work of Kodaira-Spencer and later through Mumford's theory of geometric invariants. Mumford's theory is more algebraic and deals mostly with the problem of determining whether a moduli space exists as an algebraic or projective variety, whereas the transcendental theory of Kodaira and Spencer (in fact applied in an algebraic context by Grothendieck and Artin) applies to the more general category of complex manifolds (or spaces), at the cost of producing only a local theory. In both issues, it is clear that it is not possible to have a good theory of moduli without imposing some restriction on complex manifolds or algebraic varieties.

Surfaces of general type are a case when things work out well, and one would like first to investigate properties and structure of this moduli spaces, then to draw from these results useful geometric consequences. It is my impression that for these purposes (e.g. to count number of moduli) the Kodaira-Spencer theory is by far more useful, and not difficult to apply in many concrete cases. In fact, it seems that in most applications only elementary deformation theory is needed, and that's one reason why these lecture notes cover very little of the more sophisticated theory (cf. §10 for more details). The other reason is that the author is not an expert in modern deformation theory and realized rather late about the existence or importance of some literature on the subject: in particular we would like to recommend the beautiful survey paper ([Pa]) by Palamodov on deformation of complex spaces, whose historical introduction contains rather complete information regarding the material treated in the first three lectures.

Since the style of the paper is already rather informal, we don't attempt any discussion of the main ideas here in the introduction, and, before describing with more detail the contents, we remark that the paper (according to the C. I. M. E. goals) is directed to and ought to be accessible to non specialists and to beginning graduate students. Of course, reasons of space have obliged us to assume some familiarity with the language of algebraic geometry, especially sheaves and linear systems.

Finally, in many points references are omitted for reasons of economy and the lack of a quotation of some author's name (or paper) should not be interpreted as any claim of originality on my side, or as an underestimation of some scientific work.

§1-5 summarizes the essentials of the Kodaira-Spencer-Kuranishi results needed in later sections, following existing treatments of the topic ([K-M], [Ku 3]), whereas §6 is devoted to a single but enlightening example. §7 deals with deformations of automorphisms, whereas §8-9 are devoted to Horikawa's theory of deformations of holomorphic maps, with more emphasis to applications, such as deformation of surfaces in 3-space, or of complete intersections, and include some examples of everywhere obstructed deformations, due to Mumford and Kodaira. §10 is a "mea culpa" of the author for the topics he did not treat, §11-13 try to compare Horikawa's and Schlessinger-Wahl's theory of embedded deformations, whereas §12 consists of a rewriting, with some simplifications of notation, of Kodaira's paper ([Ko 3]) treating embedded deformations of surfaces with ordinary singularities. §14-17 give a basic resumé on classification of surfaces and §18-19 are devoted to basic properties of surfaces of general type and a sketchy discussion of Gieseker's theorem on their moduli spaces. §20-23 include a rough outline of recent work of the author and a result of I. Reider: §20 deals with the number of moduli of surfaces of general type, §22 outlines the deformation theory of $(\mathbb{Z}/2)^2$ covers, §21 and 23 exhibit examples of moduli spaces with arbitrarily many connected components having different dimensions, and discuss also the problem whether the topological or the differentiable structure should be fixed.

__Acknowledgments__: It is a pleasure to thank the Centro Internazionale Matematico Estivo and the Institute for Scientific Interchange of Torino for their invitations to lecture on the topics of these notes, and for their hospitality and support. I'm also very grateful to the University of California at San Diego for hospitality and support, and especially to Ms. Annetta Whiteman for her excellent typing.

LECTURE ONE: ALMOST COMPLEX STRUCTURES and the KURANISHI FAMILY

In this lecture I will review the construction, due to Kuranishi, of the complex structures, on a compact complex manifold M, sufficiently close to the given one. To do this, one has to use the notion of almost complex structures, of integrable ones: in a sense one of the main theorems, due to Newlander and Nirenberg, is a direct extension of a basic theorem of differential geometry, the theorem of Frobenius.

§1. Almost complex structures

Let M be a differentiable (or C^{ω}, i.e. real analytic) manifold of dimension equal to $2n$, T_M its real tangent bundle.

Definition 1.1. An almost complex structure on M is the datum of a splitting $T_M \otimes \mathbb{C} = T^{1,0} \oplus T^{0,1}$, with $T^{1,0} = \overline{T^{0,1}}$.

Naturally, the splitting of $T_M \otimes \mathbb{C}$ induces a splitting for the complexified cotangent bundle $\check{T}_M \otimes \mathbb{C} = (T^{1,0})^{\vee} \oplus (T^{0,1})^{\vee}$ ($(T^{1,0})^{\vee}$ is the annihilator of $T^{0,1}$), and for all the other tensors. In particular for the rth exterior power of the cotangent bundle, one has the decomposition $\Lambda^r(\check{T}_M \otimes \mathbb{C}) = \bigoplus_{p+q=r} \Lambda^p(T^{1,0})^{\vee} \otimes \Lambda^q(T^{0,1})^{\vee}$.

We shall denote by $\mathcal{E}^{p,q}$ the sheaf of C^{∞} sections of $\Lambda^p(T^{1,0})^{\vee} \otimes \Lambda^q(T^{0,1})^{\vee}$ (resp. by $\mathcal{A}^{p,q}$ the sheaf of C^{ω} sections), by $\mathcal{E}^r$ the sheaf of C^{∞} sections of $\Lambda^r(\check{T}_M \otimes \mathbb{C})$.

The De Rham algebra is the differential graded algebra $(\mathcal{E}^*, d)$, where $\mathcal{E}^* = \bigoplus_{r=0}^{2n} \mathcal{E}^r$, and d is the operator of exterior differentiation. For a function f, $df \in \mathcal{E}^{1,0} \oplus \mathcal{E}^{0,1}$ and one can write accordingly $df = \partial f + \bar{\partial} f$; the problem is whether for all forms φ one can write $d = \partial + \bar{\partial}$, with $\partial: \mathcal{E}^{p,q} \to \mathcal{E}^{p+1,q}$, $\bar{\partial}: \mathcal{E}^{p,q} \to \mathcal{E}^{p,q+1}$ (then one has $\partial^2 = \bar{\partial}^2 = \partial\bar{\partial} + \bar{\partial}\partial = 0$, since $d^2 = 0$). Hence one poses the following

Definition 1.2. The given almost complex structure is integrable if

$$d(\mathcal{E}^{p,q}) \subset \mathcal{E}^{p+1,q} \oplus \mathcal{E}^{p,q+1} .$$

As a matter of fact, it is enough to verify this condition only for $p = 1$, $q = 0$.

Lemma 1.3. The almost complex structure is integrable $\Longleftrightarrow$ $d(\mathcal{E}^{1,0}) \subset \mathcal{E}^{2,0} \oplus \mathcal{E}^{1,1}$. [Hence another equivalent condition is: $\mathcal{E}^{1,0}$ generates a differential ideal.]

Proof. The question being local, we can take a local frame for $\mathcal{E}^{1,0}$, i.e. sections $\omega_1, \ldots, \omega_n$ of $\mathcal{E}^{1,0}$ whose values are linearly independent at each point (locally, $\mathcal{E}^{1,0}$ is a free module of rank n over $\mathcal{E}^0$, and $\{\omega_1, \ldots, \omega_n\}$ is a basis). Our weaker condition is thus that

$$(1.4) \qquad d\omega_\alpha = \sum_{\beta<\gamma} \varphi_{\alpha\beta\gamma}\, \omega_\beta \wedge \omega_\gamma + \sum_{\beta,\gamma} \psi_{\alpha\beta\bar{\gamma}}\, \omega_\beta \wedge \overline{\omega_\gamma}$$

(where $\varphi_{\alpha\gamma\beta}$ and $\psi_{\alpha\beta\bar{\gamma}}$ are functions) since every $\omega \in \mathcal{E}^{1,0}$ can be written as $\Sigma_{\alpha=1}^n f_\alpha \omega_\alpha$, and $\{\omega_\beta \wedge \omega_\gamma \mid 1 \le \beta < \gamma \le n\}$ is a local frame for $\mathcal{E}^{2,0}$, $\{\omega_\beta \wedge \bar{\omega}_\gamma \mid 1 \le \beta, \gamma \le n\}$ is a local frame for $\mathcal{E}^{1,1}$. Now $\mathcal{E}^{0,1} = \overline{\mathcal{E}^{1,0}}$ hence $d(\mathcal{E}^{0,1}) \subset \mathcal{E}^{1,1} \oplus \mathcal{E}^{0,2}$ and one verifies $d(\mathcal{E}^{p,q}) \subset \mathcal{E}^{p+1,q} \oplus \mathcal{E}^{p,q+1}$ by induction on p, q, since locally any $\eta \in \mathcal{E}^{p,q}$ can be written as $\Sigma_{\alpha=1}^n \eta_\alpha \wedge \omega_\alpha + \Sigma_{\alpha=1}^n \vartheta_\alpha \wedge \bar{\omega}_\alpha$, with $\eta_\alpha \in \mathcal{E}^{p-1,q}$, $\vartheta_\alpha \in \mathcal{E}^{p,q-1}$. Q.E.D.

At this stage, one has to observe that if M is a complex manifold, then $(T^\vee)^{1,0} = (T^{1,0})^\vee$ is generated (by definition !) by the differentials df of holomorphic functions (at least locally, if one has a chart $(z_1, \ldots, z_n)\colon U \to \mathbb{C}^n$, $dz_1, \ldots, dz_n$ give a frame for $(T^\vee)^{1,0}$). Conversely, one defines, given an almost complex structure, a function f to be holomorphic if $\bar\partial f = 0$ (i.e., $df \in \mathcal{E}^{1,0}$); one sees easily, by the local inversion theorem of U. Dini, that the almost complex structure comes from a complex structure on M if and only if for each p in M there do exist holomorphic functions $F_1, \ldots, F_n$ defined in a neighborhood U of p and giving a frame of $\mathcal{E}^{1,0}$ over U. This occurs exactly if and only if the almost complex structure is integrable: we have thus the following (cf. [N-N], [Hör] for a proof).

Theorem 1.4 (Newlander-Nirenberg). An almost complex structure on a C^∞ manifold comes from a (unique) complex structure if and only if it is integrable.

Following Weil ([We], p. 36-37) we shall give a proof in the case where everything is real-analytic, because then we see why this is an extension of the theorem of Frobenius that we now recall (see [Spiv I] for more details, or [Hi]).

Theorem 1.5. Let $\varphi_1, \ldots, \varphi_r$ be 1-forms defined in an open set Ω in $\mathbb{R}^n$ and linearly independent at any point of Ω. Then for each point p in Ω there do exist local coordinates $x_1, \ldots, x_n$ such that the span of $\varphi_1, \ldots, \varphi_r$ equals the span of $dx_1, \ldots, dx_r$, $\Longleftrightarrow$ $\varphi_1, \ldots, \varphi_r$ span a differential ideal (i.e., $\forall\, i = 1, \ldots, r$ $\exists$ forms ϑ_{ij} $(j = 1, \ldots, r)$, s.t. $d\varphi_i = \Sigma_{j=1}^r \varphi_j \wedge \vartheta_{ij}$).

Proof. The usual way to prove the theorem is to consider, $\forall\, p'$ in Ω the space $V_{p'}$ of tangent vectors killed by $\varphi_1, \ldots, \varphi_r$: then in a neighborhood U of p there exist vector fields $X_{r+1}, \ldots, X_n$ spanning $V_{p'}$ for any p' in U. Since

$$\varphi_i([X_j, X_k]) = X_j(\varphi_i(X_k)) - X_k(\varphi_i(X_j)) - d\varphi_i(X_j, X_k)$$

we see that the vector field $[X_j, X_k]$ at each p' in U lies in $V_{p'}$. One looks then for coordinates $x_1, \ldots, x_n$ s.t. $V_{p'}$ is spanned by $\partial/\partial x_{r+1}, \ldots, \partial/\partial x_n$, and these coordinates are obtained by induction on (n-r). In fact, by taking integral curves of the vector field X_n, one can assume $X_n = \partial/\partial x_n$, and replaces X_i by $Y_i = X_i - (X_i x_n) X_n$, which span the subspace $W_{p'}$ of vectors in $V_{p'}$ killing x_n, and so also the vector field $[Y_i, Y_j]$ at each point p' in U lies in $W_{p'}$ (if $X(x_n) = 0$, $Y(x_n) = 0 \Rightarrow [X,Y](x_n) = 0$!). By induction there are coordinates $(y_1, \ldots, y_n)$ with $W_{p'}$ spanned by $\partial/\partial y_{r+1}, \ldots, \partial/\partial y_{n-1}$. We can replace $X_n = \sum_{j=1}^n a_j(y)(\partial/\partial y_j)$ by $Y_n = \sum_{j=1}^r a_j(y)(\partial/\partial y_j) + a_n(y)(\partial/\partial y_n)$; since $[(\partial/\partial y_i), Y_n]$ $(i = r+1, \ldots, n-1)$ equals

$$\sum_{j \neq n+1, \ldots, n-1} \frac{\partial a_j(y)}{\partial y_i} \frac{\partial}{\partial y_j}$$

but on the other hand, this vector field is in V'_p, thus it is a multiple of Y_n by a function f. But then, on the one hand, $[(\partial/\partial y_i), Y_n](x_n) = 0$ (since $Y_n(x_n) = X_n(x_n) = 1$!), on the other hand this quantity must equal $fY_n(x_n) = f$. Hence the functions $a_j(y)$ $(j = 1, \ldots, r, n)$ depend only upon the variables $y_1, \ldots, y_r, y_n$, so, by taking integral curves of the vector field Y_n, we can assume $Y_n = \partial/\partial y_n$ also.

Q.E.D.

We have given a proof of the well known theorem of Frobenius just to notice that the only fact that is repeatedly used is the following: if X is a non zero vector field, then there exist coordinates $(x_1, \ldots, x_n)$ s.t. $X = \partial/\partial x_n$. This follows from the theorem of existence and unicity for ordinary differential equations and from Dini's theorem. Both these results hold for holomorphic functions (they are even simpler, then), therefore, given a non zero holomorphic vector field $Z = \sum_{i=1}^n a_i(w)\, \partial/\partial w_i$ on an open set in $\mathbb{C}^n$ (i.e., the a_i's are holomorphic functions), there exist local holomorphic coordinates $z_1, \ldots, z_n$ around each point such that $Z = \partial/\partial z_n$.

The conclusion is that the theorem of Frobenius holds verbatim if we replace $\mathbb{R}^n$ by $\mathbb{C}^n$, we consider holomorphic (1,0) forms $\varphi_1, \ldots, \varphi_r$, and we require local holomorphic coordinates $z_1, \ldots, z_n$ s.t. the $\mathbb{C}$-span of $\varphi_1, \ldots, \varphi_r$ be the $\mathbb{C}$-span of $dz_1, \ldots, dz_r$. The proof of the Newlander-Nirenberg theorem in the real analytic case follows then from the following.

Lemma 1.6. Let Ω be an open set in $\mathbb{R}^{2n}$, let $\omega_1,\dots,\omega_n$ be real analytic complex valued 1-forms defining an integrable almost complex structure (i.e., 1.4 holds). Then, around each point $p \in \Omega$, there are complex valued functions $F_1,\dots,F_n$ s.t. the span of $dF_1,\dots,dF_n$ equals the span of $\omega_1,\dots,\omega_n$.

Proof. Take local coordinates $x_1,\dots,x_n$ around p s.t. each ω_α is expressed by a power series $\sum_{j=1}^{2n}\sum_K f_{\alpha j,K}\, x^K dx_j$, where $K=(k_1,\dots,k_{2n})$ denotes a multi-index. Then $\overline{\omega_\alpha} = \sum_{j,K} \overline{f_{\alpha j,K}}\, x^K dx_j$ and, if we consider $\mathbb{R}^{2n}$ as contained in $\mathbb{C}^{2n}$, upon replacing the monomial x^K by the monomial z^K and x_j by dz_j (here x_j is the real part of z_j!), ω_α and $\overline{\omega_\alpha}$ extend to holomorphic 1-forms ω_α, η_α in a neighborhood of p in $\mathbb{C}^{2n}$. Since $\omega_1,\dots,\omega_n,\overline{\omega}_1,\dots,\overline{\omega}_n$ are a local frame for $\mathcal{E}^1$, the ω_α's, η_α's give a basis for the module of holomorphic 1-forms, therefore one can write

$$d\omega_\alpha = \sum_{\beta<\gamma} \varphi_{\alpha\beta\gamma}\, \omega_\beta \wedge \omega_\gamma + \sum_{\beta,\gamma} \psi_{\alpha\beta\gamma}\, \omega_\beta \wedge \eta_\gamma + \sum_{\beta<\gamma} \xi_{\alpha\beta\gamma}\, \eta_\beta \wedge \eta_\gamma \, .$$

By restriction to $\mathbb{R}^{2n}$, using (1.4) we see that $\xi_{\alpha\beta\gamma} \equiv 0$, hence $\omega_1,\dots,\omega_n$ span a differential ideal, hence Frobenius applies and there exist new holomorphic coordinates in $\mathbb{C}^{2n}$, $w_1,\dots,w_{2n}$ s.t. the span of $dw_1,\dots,dw_n$ equals the span of $\omega_1,\dots,\omega_n$. We simply take F_i to be the restriction of w_i to $\mathbb{R}^{2n}$. Q.E.D.

Remark 1.7. Assume that for $t=(t_1,\dots,t_m)$ in a neighborhood of the origin in $\mathbb{C}^m$ one is given real analytic 1-forms $\omega_{t,1},\dots,\omega_{t,n}$ as in lemma 1.6 which are expressed by convergent power series in $t_1,\dots,t_m$, and define an integrable almost complex structure when t belongs to a complex analytic subspace B containing the origin. Then, for t in B, the conclusions of lemma 1.6 hold with $F_{t,1},\dots,F_{t,n}$ expressed as convergent power series in $(t_1,\dots,t_m)$. In fact, if a vector field X_t is given by a convergent power series in $t_1,\dots,t_m$ also the solutions of the associated differential equation are power series in $t_1,\dots,t_m$: moreover, by the local inversion theorem for holomorphic functions, if $f(x,t)\colon \Omega \to \Omega$ is locally invertible, real analytic in x and complex analytic in t, then the local inverse is also complex analytic in t.

§2. Small deformations of a complex structure

If U is a vector subspace of a vector space V, and W is a supplementary subspace of U in V (thus we identify V with $U \oplus W$), then all the subspaces U', of the same dimension, sufficiently close to U, can be viewed as graphs of a linear map from U to W: we apply this principle pointwise to define a small variation of an almost complex structure (hence also of a complex structure).

Definition 2.1. A small variation of an almost complex structure is a section φ of $T^{1,0} \otimes (T^{0,1})^{\vee}$ (the variation is said to be of class C^r if φ is of class C^r).

Remark 2.2. To a small variation φ we associate the new almost complex structure s.t. $T_{\varphi}^{0,1} = \{(u,v) \in T^{1,0} \oplus T^{0,1} \mid u = \varphi(v)\}$, since there is a canonical isomorphism of $T^{1,0} \otimes (T^{0,1})^{\vee}$ with $\mathrm{Hom}(T^{0,1}, T^{1,0})$.

We assume from now on that M is a complex manifold: then, in terms of local holomorphic coordinates $(z_1, \ldots, z_n)$ one can write φ as

(2.3) $$\varphi = \sum_{\alpha,\beta} \varphi_{\alpha}^{\bar\beta}(z)\, d\bar z_{\beta} \otimes \frac{\partial}{\partial z_{\alpha}}$$

so that

$$T_{\varphi}^{0,1} = \left\{ \left(\sum_{\alpha} u_{\alpha} \frac{\partial}{\partial z_{\alpha}}, \sum_{\beta} v_{\beta} \frac{\partial}{\partial z_{\beta}} \right) \,\middle|\, u_{\alpha} = \sum_{\beta} \varphi_{\alpha}^{\bar\beta} v_{\beta} \right\}$$

and is annihilated by $(T_{\varphi}^{1,0})^{\vee}$, the span of $\{\omega_{\alpha} = dz_{\alpha} - \Sigma_{\beta} \varphi_{\alpha}^{\bar\beta} d\bar z_{\beta}\}$. On the other hand, by what we've seen $T_{\varphi}^{0,1}$ is spanned by the ξ_{γ}'s, where

$$\xi_{\gamma} = \frac{\partial}{\partial \bar z_{\gamma}} + \sum_{\alpha} \varphi_{\alpha}^{\bar\gamma} \frac{\partial}{\partial z_{\alpha}} .$$

Since $d\omega_{\alpha} = -\Sigma_{\beta}\, d\varphi_{\alpha}^{\bar\beta} \wedge d\bar z_{\beta}$, we are going to write down the integrability condition (1.4), which can be interpreted as

(2.4) $$d\omega_{\alpha}(\xi_{\gamma}, \xi_{\delta}) = 0 \quad \forall \quad \alpha, \gamma, \delta\ (\gamma < \delta) .$$

We have

$$-d\omega_{\alpha} = \sum_{\beta,\varepsilon} \left(\frac{\partial \varphi_{\alpha}^{\bar\beta}}{\partial z_{\varepsilon}} dz_{\varepsilon} \wedge d\bar z_{\beta} + \frac{\partial \varphi_{\alpha}^{\bar\beta}}{\partial \bar z_{\varepsilon}} d\bar z_{\varepsilon} \wedge d\bar z_{\beta} \right) ,$$

which belongs to $\mathcal{E}^{1,1} \oplus \mathcal{E}^{0,2}$, hence kills pairs of vectors of type (1,0). We get thus the condition

$$d\omega_{\alpha}\left(\frac{\partial}{\partial \bar z_{\gamma}}, \frac{\partial}{\partial \bar z_{\delta}}\right) + d\omega_{\alpha}\left(\sum_{\alpha'} \varphi_{\alpha'}^{\bar\gamma} \frac{\partial}{\partial z_{\alpha'}}, \frac{\partial}{\partial \bar z_{\delta}}\right)$$
$$+ d\omega_{\alpha}\left(\frac{\partial}{\partial \bar z_{\gamma}}, \sum_{\alpha''} \varphi_{\alpha''}^{\bar\delta} \frac{\partial}{\partial z_{\alpha''}}\right) = 0 ,$$

boiling down to

(2.5′) $$\frac{\partial\varphi_\alpha^{\bar\delta}}{\partial\bar z_\gamma} - \frac{\partial\varphi_\alpha^{\bar\gamma}}{\partial\bar z_\delta} + \sum_\epsilon \frac{\partial\varphi_\alpha^{\bar\delta}}{\partial z_\epsilon}\varphi_\epsilon^{\bar\gamma} - \frac{\partial\varphi_\alpha^{\bar\gamma}}{\partial z_\epsilon}\varphi_\epsilon^{\bar\delta} = 0 .$$

The condition that (2.5′) holds for each α, and $\gamma < \delta$, can be written more simply as

(2.5) $$\bar\partial\varphi = \frac{1}{2}[\varphi,\varphi] ,$$

where

$$\bar\partial\varphi = \sum_\alpha \left(\sum_{\gamma<\delta} \left(\frac{\partial\varphi_\alpha^{\bar\gamma}}{\partial\bar z_\delta} - \frac{\partial\varphi_\alpha^{\bar\delta}}{\partial\bar z_\gamma} \right) d\bar z_\gamma \wedge d\bar z_\delta \right) \otimes \frac{\partial}{\partial z_\alpha} ,$$

$$[\varphi,\varphi] = 2 \sum_\alpha \sum_{\gamma<\delta} \left(\sum_\epsilon \frac{\partial\varphi_\alpha^{\bar\delta}}{\partial z_\epsilon}\varphi_\epsilon^{\bar\gamma} - \frac{\partial\varphi_\alpha^{\bar\gamma}}{\partial z_\epsilon}\varphi_\epsilon^{\bar\delta} \right) d\bar z_\gamma \wedge d\bar z_\delta \otimes \frac{\partial}{\partial z_\alpha}$$

$$= \sum_{\alpha,\epsilon,\gamma,\delta} d\bar z_\gamma\, \varphi_\epsilon^{\bar\gamma} \left(\frac{\partial\varphi_\alpha^{\bar\delta}}{\partial z_\epsilon} \right) \wedge d\bar z_\delta \otimes \frac{\partial}{\partial z_\alpha} - d\bar z_\delta\, \varphi_\alpha^{\bar\delta} \left(\frac{\partial\varphi_\epsilon^{\bar\gamma}}{\partial z_\alpha} \right) \wedge d\bar z_\gamma \otimes \frac{\partial}{\partial z_\epsilon} .$$

We shall explain these definitions while recalling some standard facts on Dolbeault cohomology and Hodge theory (harmonic forms).

So, let V be a holomorphic vector bundle, and let (U_α) be a cover of M by open sets where one has a trivialization $V_{|U_\alpha} \cong U_\alpha \times \mathbb{C}^r$, hence fibre vector coordinates v_α, related by $v_\alpha = g_{\alpha\beta} v_\beta$ where $g_{\alpha\beta}$ is an invertible $r \times r$ matrix of holomorphic functions. We let $\mathcal{E}^{0,p}(V)$ be the space of (C^∞) sections of $V \otimes \Lambda^p(T^{0,1})^V$: since $\bar\partial g_{\alpha\beta} = 0$, it makes sense to take $\bar\partial$ of (0, p) forms with values in V (i.e., elements of $\mathcal{E}^{0,p}(V)$), and we have the Dolbeault exact sequence of sheaves

$$0 \to \mathcal{O}(V) \to \mathcal{E}(V) \xrightarrow{\bar\partial_1} \mathcal{E}^{0,1}(V) \xrightarrow{\bar\partial_2} \cdots \xrightarrow{\bar\partial_n} \mathcal{E}^{0,n}(V) \to 0 ,$$

where $\mathcal{O}(V)$ is the sheaf of holomorphic sections of V. We have the theorem of Dolbeault (the $\mathcal{E}^{0,k}(V)$ are soft sheaves).

<u>Theorem 2.6.</u>

$$H^i(M, \mathcal{O}(V)) \cong \frac{\ker H^0(\bar\partial_{i+1})}{\operatorname{Im} H^0(\bar\partial_i)} .$$

So $\bar\partial$ is well defined for our $\varphi \in \mathcal{E}^{0,1}(T^{1,0})$. For further use, we shall use the notation $\Theta = \mathcal{O}(T^{1,0})$. To explain the bracket operation, we notice that this is a bilinear operation

$$[\ ,\ ]: \mathcal{E}^{o,p}(T^{1,0}) \times \mathcal{E}^{o,q}(T^{1,0}) \to \mathcal{E}^{o,p+q}(T^{1,0})$$

which in local coordinates $(z_1, \ldots, z_n)$, if

$$\varphi = \sum_{I=\{i_1<\cdots<i_p\}} \sum_\alpha f_\alpha^{\bar I}(d\bar z_{i_1} \wedge \ldots \wedge d\bar z_{i_p}) \otimes \frac{\partial}{\partial z_\alpha} = \sum_{I,\alpha} f_\alpha^{\bar I}\, d\bar z^{\wedge I} \otimes \frac{\partial}{\partial z_\alpha}$$

and

$$\psi = \sum_{J,\varepsilon} g_\varepsilon^{\bar J}\, d\bar z^{\wedge \bar I} \otimes \frac{\partial}{\partial z_\varepsilon} \quad ,$$

is such that

$$[\varphi, \psi] = \sum_{I,J,\alpha,\varepsilon} d\bar z^{\wedge I} \wedge d\bar z^{\wedge J} \otimes \left[f_\alpha^{\bar I} \frac{\partial g_\varepsilon^{\bar J}}{\partial z_\alpha} \frac{\partial}{\partial z_\varepsilon} - g_\varepsilon^{\bar J} \frac{\partial f_\alpha^{\bar I}}{\partial z_\varepsilon} \frac{\partial}{\partial z_\alpha} \right]$$

The bracket operation enjoys the following properties

i) $[\psi, \varphi] = (-1)^{pq+1}[\varphi, \psi]$

(2.7) ii) $\bar\partial[\varphi, \psi] = [\bar\partial\varphi, \psi] + (-1)^p[\varphi, \bar\partial\psi]$

iii) if ξ is in $\mathcal{E}^{o,r}(T^{1,0})$, then the Jacobi identity holds, i.e.,

$$(-1)^{pr}[\varphi,[\psi,\xi]] + (-1)^{qp}[\psi,[\xi,\varphi]] + (-1)^{rq}[\xi,[\varphi,\psi]] = 0 .$$

Before recalling the Hodge theory of harmonic forms, we remark that, if we have a small variation $\varphi(t)$ of complex structure depending on a parameter $t = (t_1, \ldots, t_m)$, setting $B = \{t \mid \bar\partial\varphi(t) = \frac{1}{2}[\varphi(t), \varphi(t)]\}$, B is precisely the set of points t for which $\varphi(t)$ defines a complex structure: but in order that the complex charts depend holomorphically upon t for t in B (we assume, of course, that $\varphi(t)$ be a power series in $t_1, \ldots, t_m$), we want (cf. remark 1.6) B to be a complex subspace. The Kuranishi family, as will be explained in the second lecture, is a natural choice to embody all the small variations of complex structures with the smallest number of parameters.

Now, let V be again a holomorphic vector bundle on M, and assume that we choose Hermitian metrics for V and $T^{1,0}$, so that for all the bundles $V \otimes (T^{o,p})^\vee$ is determined a Hermitian metric (if M is C^ω, we can assume the metric to be C^ω). Thus a volume form $d\mu$ is given also on M, and thus, for $\varphi, \psi \in \mathcal{E}^{o,p}(V)$ a Hermitian scalar product is defined by $(\varphi, \psi) = \int_M \langle \varphi, \psi \rangle_x \, d\mu$. ($\langle \varphi, \psi \rangle_x$ is the value which the Hermitian product, given for the fibre of $V \otimes (T^{o,p})^\vee$ at the point x, takes on the values of φ and ψ at x).

It is therefore defined the adjoint operator $\bar\partial^*: \mathcal{E}^{o,p+1}(V) \to \mathcal{E}^{o,p}(V)$ by the usual formula $(\bar\partial\varphi, \xi) = (\varphi, \bar\partial^* \xi)$, and one forms the Laplace operator

$$\Box = \bar\partial\,\bar\partial^* + \bar\partial^*\,\bar\partial\,.$$

We have $\Box: \mathcal{E}^{o,p}(V) \to \mathcal{E}^{o,p}(V)$ and the space of harmonic forms is

$$\mathcal{H}^p(V) = \{\varphi \in \mathcal{E}^{o,p}(V) \mid \Box\varphi = 0\} = \{\varphi \mid \partial\varphi = \bar\partial^*\varphi = 0\} \tag{2.8}$$

The main result is that one has an orthogonal direct sum decomposition (where we simply write $\mathcal{E}^p$ for $\mathcal{E}^{o,p}(V)$)

$$\mathcal{E}^p = \mathcal{H}^p \overset{\perp}{\oplus} \bar\partial\mathcal{E}^{p-1} \overset{\perp}{\oplus} \bar\partial^*\,\mathcal{E}^{p+1} \tag{2.9}$$

Remark 2.10. $\mathcal{H}^p \overset{\perp}{\oplus} \bar\partial\,\mathcal{E}^{p-1}$ consists of the space $\Gamma(\ker\bar\partial)$ of all the $\bar\partial$ closed p-forms: in fact if $\bar\partial\,\bar\partial^*\varphi = 0$, then $0 = (\bar\partial\,\bar\partial^*\varphi, \varphi) = \|\bar\partial^*\varphi\|^2 \Rightarrow \bar\partial^*\varphi = 0$. Therefore, in view of Dolbeault's theorem one has the following

Theorem 2.11 (Hodge). $\mathcal{H}^p(V)$ is naturally isomorphic to $H^p(M, \mathcal{O}(V))$. Moreover for each $\varphi \in \mathcal{E}^{o,p}(V)$ there is a unique decomposition

$$\varphi = \eta + \Box\psi\,, \quad \text{with} \quad \eta = H(\varphi) \in \mathcal{H}^p\,, \ \psi = G(\varphi) \in (\mathcal{H}^p)^\perp\,.$$

H is obviously a projector (the "harmonic projector") onto the finite dimensional space $\mathcal{H}^p$, whereas G is called the Green operator. We refer to [K-M] again for the proof of the following

Proposition 2.12. $\bar\partial$, $\bar\partial^*$ commute with G, and the product of $\bar\partial$, $\bar\partial^*$ or G with H on both sides gives zero.

§3. Kuranishi's equation and the Kuranishi family

Fix once for all an Hermitian metric on $T^{1,0}$ and let $\mathcal{H}^p$ be $\mathcal{H}^p(T^{1,0})$: we can therefore identify, by Hodge's theorem, harmonic forms in $\mathcal{H}^p$ with cohomology classes in $H^p(M, \Theta)$. Recall also that, by (2.7.ii), a bracket operation is defined

$$[\ ]: H^p(M, \Theta) \times H^q(M, \Theta) \to H^{p+q}(M, \Theta).$$

Let $\eta_1, \dots, \eta_m$ be a basis for $\mathcal{H}^1$, so that we can identify a point $t \in \mathbb{C}^m$ with the harmonic form $\sum_{i=1}^m t_i \eta_i$. Consider the following equation

$$\varphi(t) = \sum t_i \eta_i + \frac{1}{2}\bar\partial^* G[\varphi(t), \varphi(t)]\,. \tag{3.1}$$

It is easy to see that one has a formal power series solution $\varphi = \sum_{m=1}^{\infty} \varphi_m(t)$, where $\varphi_m(t)$ is homogeneous of degree m in t: in fact by linearity on t of $\bar\partial^*$, G

$$\varphi_1(t) = \sum t_i \eta_i \ , \quad \varphi_2(t) = \frac{1}{2}\,\bar\partial^* G[\varphi_1(t), \varphi_1(t)], \quad \varphi_3 = \bar\partial^* G[\varphi_1, \varphi_2], \ \ldots$$

The power series converges in a neighborhood of the origin because G is a regularizing operator of order 2 (with respect to Hölder or Sobolev norms).

We want to show that $B = \{ t \mid \varphi(t)$ converges, and defines a complex structure on $M \}$ is a complex subspace around the origin in $\mathbb{C}^m$. We know that $B = \{ t \mid \bar\partial\varphi(t) - \frac{1}{2}[\varphi(t), \varphi(t)] = 0 \}$ and we claim that the following holds

Lemma 3.2. $\bar\partial\varphi(t) - \frac{1}{2}[\varphi(t), \varphi(t)] = 0$ if and only if $H[\varphi(t), \varphi(t)] = 0$.

Proof. The "only if" part is clear, since $H\bar\partial = 0$. Conversely, we want to show that $\psi = \bar\partial\varphi - \frac{1}{2}[\varphi, \varphi]$ equals zero. Now $\varphi = \varphi_1 + \frac{1}{2}\bar\partial^* G[\varphi, \varphi]$ by Kuranishi's equation, and φ_1 is harmonic: hence

$$\psi = \frac{1}{2}\,\bar\partial\bar\partial^* G[\varphi, \varphi] - \frac{1}{2}[\varphi, \varphi] = \frac{1}{2}\,(\bar\partial\bar\partial^* G - \mathrm{id})([\varphi, \varphi]).$$

But the identity id equals $H + \Box G = H + \bar\partial\bar\partial^* G + \bar\partial^*\bar\partial G$, thus (since $H[\varphi, \varphi] = 0$ by assumption) $-2\psi = \bar\partial^*\bar\partial G[\varphi, \varphi] =$ (since $\bar\partial$, G commute) $= \bar\partial^* G\ \bar\partial[\varphi, \varphi] =$ (by 2.7) $= 2\bar\partial^* G[\bar\partial\varphi, \varphi] =$ (since $[[\varphi, \varphi], \varphi] = 0$ by Jacobi's identity) $= 2\bar\partial^* G[\psi, \varphi]$. We have therefore reached the conclusion that $\psi(t) = -\bar\partial^* G[\psi(t), \varphi(t)]$, in particular for any Sobolev norm $\| \ \|$, $\|\psi(t)\| \le \text{cost.}\ \|\psi(t)\|\ \|\varphi(t)\|$. But since $\|\varphi(t)\|$ is infinitesimal as $t \to 0$, we get that for t small $\|\psi(t)\| < \|\psi(t)\|$, hence $\|\psi(t)\| = 0$ and $\psi(t) = 0$ as we want to show. Q.E.D.

We use now the standard notation $h^i(\mathcal{F}) = \dim_{\mathbb{C}} H^i(M, \mathcal{F})$, for a coherent sheaf on M, and we state an immediate consequence of (3.2).

Corollary 3.3. If $m = h^1(\Theta)$ as before, $k = h^2(\Theta)$, then B is defined by k holomorphic functions $g_1, \ldots, g_k$ of $t = (t_1, \ldots, t_m)$ which have multiplicity at least 2 at the origin. Moreover, if we identify $\mathbb{C}^m$ with $H^1(\Theta)$, $\mathbb{C}^k$ with $H^2(\Theta)$, the function $g^{(2)}: \mathbb{C}^m \to \mathbb{C}^k$ given by the quadratic terms of $g_1, \ldots, g_k$ corresponds to the quadratic function associated to the symmetric bilinear function

$$[\ ,\ ]: H^1(\Theta) \times H^1(\Theta) \to H^2(\Theta).$$

Proof. Let $\xi_1, \ldots, \xi_k$ be an orthonormal basis for $\mathcal{H}^2$. If ω is in $\mathcal{E}^{0,2}(T^{1,0})$, $H\omega = 0$ is equivalent to $(\omega, \xi_i) = 0$ for $i = 1, \ldots, k$. Therefore, by lemma 3.2, B is defined by the k functions $g_i(t) = ([\varphi(t), \varphi(t)], \xi_i) = 0$. Since

$\varphi(t) = \sum t_i \eta_i + o(t)$, $g_i(t)$, which is clearly a convergent power series in t, has a McLaurin expansion

$$g_i(t) = \sum_{j,k=1}^{m} ([\eta_j, \eta_k], \xi_i) t_j t_k + o(t^2). \qquad \text{Q.E.D.}$$

The final step is to observe that the varying complex structures on M, parametrized by $t \in B$, can be put together to give a structure of complex space to the product $M \times B$, we have

Theorem 3.4. On $M \times B$ there exists a structure of complex space $\mathcal{X}$ such that the projection on the second factor induces a holomorphic map $\Pi: \mathcal{X} \to B$ such that

i) each fiber $X_t = \Pi^{-1}(t)$ is the complex manifold obtained by endowing M with the complex structure defined by $\varphi(t)$;

ii) for each point $p \in M = X_0$ there exists a neighborhood U in M and a neighborhood V in $\mathcal{X}$ such that V is biholomorphic to $U \times B$ under a map $\psi: U \times B \to V$ s.t. $\pi \circ \psi$ is projection on the second factor.

Sketch of Proof. By remark 1.7, for each point p in M there is a neighborhood U and functions $F_{t,i}(x)$ $(i = 1, \dots, n)$ s.t. for any t in B they give a local chart for the complex structure defined by $\varphi(t)$. Let $F(t, x) = (F_{t,1}(x), \dots, F_{t,n}(x))$: $U \to \mathbb{C}^n$; we use (F, t): $U \times B \to \mathbb{C}^n \times B$ to give the local charts for the complex structure $\mathcal{X}$. The inversion theorem of U. Dini ensures then that the complex structure on $\mathcal{X}$ is globally well defined, and that ii) holds.

Q.E.D.

LECTURE TWO: DEFORMATIONS OF COMPLEX STRUCTURES AND KURANISHI'S THEOREM

In this lecture I will review the notion of deformation of complex structure introduced by Kodaira and Spencer, the notion of pull-back, versal family,..., define the Kodaira-Spencer map, and state the theorem about the semi-universality of the Kuranishi family.

§4. Deformations of complex structure

Let M be as usual a compact complex manifold.

Definition 4.1. A deformation of M consists of the following data: a morphism of complex spaces $\Pi: \mathcal{X} \to B$, a point $0 \in B$, an isomorphism of the fiber $X_0 = \Pi^{-1}(0)$ with M s.t. Π is proper and flat.

Remark 4.2. Π is said to be smooth if $\forall\, b \in B$ the fibre $X_b = \Pi^{-1}(b)$ is smooth (and reduced, of course!). A deformation Π is smooth (at least if one shrinks B) by virtue of the following

Lemma 4.3. Let $\Pi^*: \mathcal{O}_{B,o} = A \to \mathcal{O}_{\mathcal{X},p} = R$ be a homomorphism of local rings of complex spaces, and assume that Π^* makes R a flat A-module, and that moreover Π^* has a smooth fibre, i.e. $R/\Pi^*\mathfrak{m}_A \cong \mathbb{C}\{\bar{x}_1, \ldots, \bar{x}_n\}$, $\mathfrak{m}_A$ being the maximal ideal of A. Then $R \cong A\{x_1, \ldots, x_n\}$.

Proof. Let $x_1, \ldots, x_n$ be such that x_i maps to $\bar{x}_i$ through the surjection $R \twoheadrightarrow R/\Pi^*\mathfrak{m}_A$. Thus Π^* defines a homomorphism $f: A\{x_1, \ldots, x_n\}$ to R. f is surjective by Nakayama's lemma, and we claim that flatness implies the injectivity of f. Let $K = \ker f$, so that we have an exact sequence

$$0 \to K \to A\{x_1, \ldots, x_n\} \to R \to 0 .$$

Tensoring with the A module $A/\mathfrak{m}_A \cong \mathbb{C}$ we get, since $\mathrm{Tor}_1^*(R, A/\mathfrak{m}_A) = 0$ (see [Dou 2], proposition 3) an exact sequence

$$0 \to K \otimes A/\mathfrak{m}_A \to \mathbb{C}\{\bar{x}_1, \ldots, \bar{x}_n\} \overset{\Pi^*}{\to} R/\Pi^*\mathfrak{m}_A \to 0 .$$

Since Π^* is, by assumption, an isomorphism, $K \otimes A/\mathfrak{m}_A = 0$, hence $K = 0$, again by Nakayama's lemma (applied to K as an R module!).

Q.E.D.

Remark 4.4. A deformation is said to be smooth if B is smooth: in this case Π is just a proper map with surjective differential at each point. Lemma 4.3 shows that property ii) of theorem 3.4 holds for every deformation. In the case when B is smooth, a classical theorem of Ehresmann ([Eh]) asserts that Π is a differentiable fibre bundle. This is a local result, and to give an idea of the proof (especially to stress the importance of vector fields!), we can assume B to be a k cube in $\mathbb{R}^k$. Then one wants to show that $\mathcal{X}$ is diffeomorphic to $M \times B$ vis a map compatible with the two projections on B: one assumes $\mathcal{X}$ to have a Riemannian metric, so that, if $x_1, \dots, x_k$ are coordinates in $\mathbb{R}^k$, one can lift the vector field $\partial/\partial x_k$ to $\mathcal{X}$ to a unique vector field ξ that is orthogonal to the fibres of Π. Then one proves the result by induction on k: if $B' = B \cap \mathbb{R}^{k-1}$, one takes the integral curves of ξ to construct a diffeomorphism of $\Pi^{-1}(B') \times (-1, 1)$ with $\mathcal{X}$, and then applies induction ($\Pi^{-1}(B') \cong M \times B'$) to infer that $\mathcal{X} \cong M \times B$. If everything (the Riemannian metric included) is C^ω, the above proof yields a C^ω diffeomorphism. An analogous result holds in the general case.

Theorem 4.5. Given a deformation $\Pi: \mathcal{X} \to B$ (shrinking B if necessary) there exists a real analytic (C^ω) diffeomorphism $\gamma: M \times B \to \mathcal{X}$ with $\Pi \circ \gamma$ = projection of $M \times B \to B$, and such that γ is holomorphic in the second set of variables.

Idea of proofs. In the C^∞ case (cf. [Ku 3], p. 19-23) the proof is easier: there exists a finite cover V_α of $\mathcal{X}$ such that, if $U_\alpha = M \cap V_\alpha$ (M is identified with X_o), $V_\alpha \cong U_\alpha \times B$ under a biholomorphism φ_α. We can assume $\mathcal{X} \subset \mathbb{R}^N$: using these φ_α's and a partition of unity subordinate to the cover V_α we can define a C^∞ morphism of $\mathcal{X}$ to a tubular neighborhood T of M in $\mathbb{R}^N$, and then we compose with a retraction of T to M to get $\varphi: \mathcal{X} \to M$ such that $\varphi_{|M}$ = identity. Then $\varphi \times \Pi$ gives the required diffeomorphism.

In the real analytic case, one can use the fact that, if T_M is the real tangent bundle of M, $H^1(M, T_M) = 0$: then the power series method of [K-M], pp. 45-55 gives the desired result.

Using the diffeomorphism $\gamma: M \times B \to \mathcal{X}$, for each $b \in B$ one gets a small variation of complex structure $\varphi(b) \in A^{0,1}(T^{1,0})$ which depends holomorphically upon b. If $(B, o) \subset (\mathbb{C}^r, o)$ and $t_1, \dots, t_r$ are coordinates on $\mathbb{C}^r$ ($r = \dim \mathfrak{m}_{B,o} / \mathfrak{m}^2_{B,o}$) one can write $\varphi(b) = \Sigma\, t_i \eta_i + o(t)$, and the linear map ρ from $\mathbb{C}^r = T_{B,o}$ (Zariski tangent space to B at 0) to $H^1(\Theta)$ such that $-\rho(\partial/\partial t_i)$ = class of η_i in $H^1(\Theta_M)$, is called the Kodaira-Spencer map, and we shall soon give an easier way to define and compute it.

Definition 4.6. Let $\Pi: \mathcal{X} \to B$ be a deformation of M, and let V_α be a (finite) cover of $\mathcal{X}$ which locally trivializes Π, i.e. such that there exists a biholomorphism $\varphi_\alpha: V_\alpha \to U_\alpha \times B$ ($U_\alpha = V_\alpha \cap M$, and we tacitly assume $\Pi \circ \varphi_\alpha^{-1}$ to be the projection of $U_\alpha \times B \to B$). Let ξ be a tangent vector to B at 0, and let ϑ_α be the unique lifting of ξ (viewed as a constant vector field in $\mathbb{C}^r \supset B$) to V_α given by the trivialization φ_α. Then $\vartheta_\alpha - \vartheta_\beta$, restricted to $U_\alpha \cap U_\beta$, is a vertical vector field, thus $(\vartheta_\alpha - \vartheta_\beta) \in H^1(M_\alpha, \Theta)$. We define $\rho: T_{B,o} \to H^1(\Theta)$ to be the linear map such that $\rho(\xi) = \vartheta_\alpha - \vartheta_\beta$ and it is easy to see that ρ is well defined, independently of the choice of the cover and of the trivializations (for instance, changing trivializations, ϑ is replaced by ϑ'_α such that $\vartheta_\alpha - \vartheta'_\alpha$ is a vertical vector field, therefore $\vartheta_\alpha - \vartheta_\beta$ is cohomologous to $\vartheta'_\alpha - \vartheta'_\beta$). ρ is called the Kodaira-Spencer map.

The Kodaira-Spencer map gives the first order obstruction to the global liftability of a vector field, and its importance lies in its functorial nature, that we are now going to explain.

Definition 4.7. Let $\Pi: \mathcal{X} \to B$ be a deformation of M, and let $f: B' \to B$ be a morphism of complex spaces, O' a point of B' with $f(O') = 0$. Then the pull back $f^*(\mathcal{X})$ is given by $\mathcal{X}' = \{(x, b') \mid x \in \mathcal{X},\ b' \in B' \text{ s.t. } \Pi(x) = f(b')\} \subset \mathcal{X} \times B'$, with Π' induced by projection on the second factor.

Let $f_*: T_{B',o'} \to T_{B,o}$ be the differential of f at O', and let ρ, ρ' be the respective Kodaira-Spencer maps of $\mathcal{X}$, $\mathcal{X}'$: we have

(4.8) $$\rho' = \rho \circ f_*$$

as it is immediately verified.

Grothendieck's point of view was, in particular, that in order to compute $\rho(\xi)$ it suffices to choose $B' = \{t \in \mathbb{C} \mid t^2 = 0\}$, and f the unique morphism of $B' \to B$ s.t. $f_*(\partial/\partial t) = \xi$; also, in the context of pull-back, the meaning of $\rho = 0$ is that, if B^I is the subspace of B defined by $\mathfrak{m}^2_{B,o}$, $i: B^I \to B$ is the inclusion, then $i^*(\mathcal{X}) \cong B^I \times M$ if and only if $\rho = 0$.

We can thus verify that the two definitions we have given of ρ do, in fact, coincide, limiting ourselves to 1-parameter deformations.

Lemma 4.9. Let $\Pi: \mathcal{X} \to B$ be a deformation of M with base $B = \{t \in \mathbb{C} \mid t^2 = 0\}$, whose associated small variation of complex structure is given by the form $\varphi(t) = t\eta$, with $\eta \in A^{0,1}(T^{1,0})$: then, using the Dolbeault isomorphism, $\rho(\partial/\partial t)$ is the class of η in $H^1(\Theta)$.

Proof. We choose trivializing charts on $U_\alpha \times B$, with $z_1^\alpha, \ldots, z_n^\alpha$ coordinates in U_α, given by $\zeta_j^\alpha(z,t) = z_j^\alpha + t w_j^\alpha$. $\varphi(t)$ on U_α is expressed by

$$t \sum_{i,j} \eta_i^{\alpha \bar{j}} \, d\bar{z}_j^\alpha \otimes \frac{\partial}{\partial z_i^\alpha}$$

and the condition that ζ_i^α be holomorphic is that $(\bar{\partial} + \varphi)(\zeta_i^\alpha) = 0$, i.e.

$$\bar{\partial} w_i^\alpha + \sum_j \eta_i^{\alpha \bar{j}} \, d\bar{z}_j^\alpha = 0 ,$$

hence locally η can be expressed as

$$- \sum_i \bar{\partial} w_i^\alpha \otimes \frac{\partial}{\partial z_i^\alpha} .$$

In view of the way the Dolbeault isomorphism is gotten, it suffices to verify that, if we set

$$- \sum_i w_i^\alpha \otimes \frac{\partial}{\partial z_i^\alpha} = \Lambda_\alpha$$

then $\Lambda_\alpha - \Lambda_\beta$ is cohomologous to $\vartheta_\alpha - \vartheta_\beta$. Now $\vartheta_\alpha = \partial/\partial t$ in the α-chart, therefore, expressing $\vartheta_\alpha - \vartheta_\beta$ in the α-chart, we get, if $\zeta_j^\alpha = f_j^{\alpha\beta}(\zeta^\beta, t)$ is the change of coordinates,

$$- \sum_i \left. \frac{\partial f_i^{\alpha\beta}(\zeta^\beta, t)}{\partial t} \right|_{t=0} \frac{\partial}{\partial z_i^\alpha} = - \sum_i g_i^{\alpha\beta}(z^\beta) \frac{\partial}{\partial z_i^\alpha}$$

if we set

$$f_j^{\alpha\beta}(\zeta^\beta, t) = h_j^{\alpha\beta}(\zeta^\beta) + t g_j^{\alpha\beta}(\zeta^\beta) .$$

But this last expression equals $\zeta_j^\alpha = z_j^\alpha + t w_j^\alpha$, hence, by the chain rule,

$$w_i^\alpha = g_i^{\alpha\beta}(z^\beta) + \sum_k \frac{\partial h_i^{\alpha\beta}}{\partial z_k^\beta} \cdot w_k^\beta ,$$

and we are done, since

$$\frac{\partial}{\partial z_k^\beta} = \sum_i \frac{\partial h_i^{\alpha\beta}}{\partial z_k^\beta} \frac{\partial}{\partial z_i^\alpha} . \qquad \text{Q.E.D.}$$

§5. Kuranishi's theorem

Definition 5.1. A deformation $\Pi: \mathcal{X} \to B$ of M is said to be complete if for any other deformation $\Pi': \mathcal{X}' \to B'$, there exists a neighborhood B'' of O' in B' and $f: B'' \to B$ such that $\mathcal{X}'' = \mathcal{X}'|B''$ is isomorphic to the pull-back $f^*(\mathcal{X})$. The deformation $\Pi: \mathcal{X} \to B$ is said to be universal if it is complete, and moreover f is (locally) unique (respectively: semi-universal if it is complete and $f_*: T_{B',o'} \to T_{B,o}$ is unique).

Remark 5.2. In view of (4.8), a complete deformation is semi-universal if the associated Kodaira-Spencer map ρ is injective. Let us see that ρ is surjective for a complete family; in fact, by lemma (4.9) and the Kuranishi equation (3.1), the Kuranishi family has a bijective Kodaira-Spencer map: hence, if a complete family exists, it must have a surjective Kodaira-Spencer map.

Proposition 5.3. A semi-universal family is unique up to isomorphism.

Proof. By completeness, if $\Pi: \mathcal{X} \to B$, $\Pi': \mathcal{X}' \to B'$ are semi-universal, there exist $f': B \to B'$, $f: B' \to B$ with $\mathcal{X}' = f^*(\mathcal{X})$, $\mathcal{X} = f'^*(\mathcal{X}')$. Hence $\mathcal{X} = (f'f)^*(\mathcal{X})$, and by semi-universality $(f'f)_* =$ identity, but also $(f f')_* =$ id, therefore, by the local inversion theorem, f, f' are isomorphisms. Q.E.D.

We can now state the theorem of Kuranishi, referring the reader, for a complete proof, to [Dou 1], [Ku 3], [Ku 2], and to [K-M] and [Ku 1] for weaker versions.

Theorem 5.4 (Kuranishi)

i) The Kuranishi family is semi-universal, and f_* coincides (up to sign) with the Kodaira-Spencer map ρ.

ii) The Kuranishi family is complete for $b \in B - \{o\}$, when viewed as a deformation of X_b.

iii) If $H^0(\Theta_M) = 0$, the Kuranishi family is universal.

Let's draw some corollaries of the above theorem, noting that by proposition 5.3, the Kuranishi family is "the" semi-universal family of deformations, and that, by (3.3), the Kuranishi family is smooth if $H^2(\Theta) = 0$.

Corollary 5.5. If a deformation $\Pi': \mathcal{X}' \to B'$ has a smooth base B', and surjective Kodaira-Spencer map ρ', then it is complete and moreover the Kuranishi family of deformations is smooth.

Proof. Let $f : B' \to B$ be such that $\mathcal{X}' = f^*(\mathcal{X})$. By taking a smooth submanifold of B', we can assume f_* to be bijective. But then B' is a neighborhood of $O \in \mathbb{C}^r$, and we have $f: \mathbb{C}^r \to B \subset \mathbb{C}^r$ with f_* invertible: by the local inversion theorem $f(\mathbb{C}^r)$ contains a neighbourhood of $O \in \mathbb{C}^r$, hence gives a local isomorphism $f: B' \cong \mathbb{C}^r \overset{\cong}{\to} B \cong \mathbb{C}^r$. Q.E.D.

Finally, we mention a refinement of part iii) of the theorem ([Wav]).

Theorem 5.6 (Wavrick). If the base B of the Kuranishi family is reduced, and $h^o(X_t, \Theta_t)$ is constant, then the Kuranishi family is universal.

In the next paragraph we shall discuss the example of the Segre-Hirzebruch surfaces, which illustrates how certain statements in the above theorems cannot be improved. We only remark here that if M is a curve ($n = 1$), the $H^2(\Theta) = 0$ and the Kuranishi family is smooth of dimension $h^1(\Theta) = 3g - 3 + a$, where g is the genus of the curve and $a = h^o(\Theta)$ is the dimension of the group of automorphisms of M.

§6. The example of Segre-Hirzebruch surfaces

The Segre-Hirzebruch surface $\mathbb{F}_n$ (where $n \in \mathbb{N}$) is, in fancy language, the $\mathbb{P}^1$ bundle $\mathbb{P}(V_n)$ associated to the rank 2 vector bundle V_n such that $\mathcal{O}(V_n) \cong \cong \mathcal{O}_{\mathbb{P}^1} \oplus \mathcal{O}_{\mathbb{P}^1}(n)$. By abuse of language we shall identify V with $\mathcal{O}(V)$, therefore we get a split exact sequence

$$(6.1) \qquad O \to \mathcal{O}_{\mathbb{P}^1} \to V_n \to \mathcal{O}_{\mathbb{P}^1}(n) \to 0 .$$

We consider all the rank 2 vector bundles V which fit into an exact squence like (6.1), which are classified by $H^1(\mathcal{O}_{\mathbb{P}^1}(-n))$, a vector space of dimension $(n-1)$, and we consider the family of ruled surfaces $\mathbb{P}(V)$, thus obtained, as a deformation of $\mathbb{F}_n$. In concrete terms, we take $B = \mathbb{C}^{n-1}$, with coordinates $t_1, \ldots, t_{n-1}$, and we obtain $\mathcal{X}$ glueing $\mathbb{P}^1 \times \mathbb{C} \times B$ $(= \mathbb{P}^1 \times (\mathbb{P}^1 - \{\infty\}) \times B)$ with $\mathbb{P}^1 \times \mathbb{C} \times B$ $(= \mathbb{P}^1 \times (\mathbb{P}^1 - \{o\}) \times B)$ by the identification of $(y_0, y_1, z, t_1, \ldots, t_{n-1})$ with $(y_0', y_1', z', t_1, \ldots, t_{n-1})$ if

$$(6.2) \qquad \boxed{z' = \frac{1}{z}, \quad y_1' = y_1 z^{-n}, \quad y_0' = y_0 + y_1 \sum_{i=1}^{n-1} t_i z^{-i}}$$

(note then that $y_0 = y_0' + y_1' \cdot \Sigma_i t_i z^{n-i}$).

Now we shall compute the Kodaira-Spencer map of the family $\Pi: \mathcal{X} \to B$ we have just constructed.

$$\rho\left(\frac{\partial}{\partial t_i}\right) = \left(\frac{\partial}{\partial t_i}\right) - \left(\frac{\partial}{\partial t_i}\right)' ,$$

where the prime means we are writing a vector field using the second chart. So, expressing $(\partial/\partial t_i)$ using the first chart, we get

$$\rho\left(\frac{\partial}{\partial t_i}\right) = -(y_1' z^{n-i}) \frac{\partial}{\partial y_0} = (-y_1 z^{-i}) \frac{\partial}{\partial y_0} .$$

We shall now show that these vector fields generate the Čech cohomology group $H^1(\{U, U'\}, \Theta_{\mathbb{F}_n})$, where $U = \mathbb{P}^1 \times \mathbb{C}$ of the second chart, and we notice that, since $H^1(\Theta_{\mathbb{P}^1 \times \mathbb{C}}) = 0$, the Čech cohomology group we are going to compute is indeed $H^1(\mathbb{F}_n, \Theta_{\mathbb{F}_n})$, hence the Kodaira-Spencer map will be bijective, and the constructed family will be the Kuranishi family.

Let's work on $\mathbb{F}_n$, where $z' = 1/z$, $y_1' = y_1 z^{-n}$, $y_0' = y_0$: since on $\mathbb{P}^m$ we have the Euler exact sequence

(6.3) $$0 \to \Theta_{\mathbb{P}^m} \xrightarrow{\begin{pmatrix} x_0 \\ \vdots \\ x_m \end{pmatrix}} \Theta_{\mathbb{P}^m}(1)^{m+1} \xrightarrow{\left(\frac{\partial}{\partial x_0}, \dots, \frac{\partial}{\partial x_m}\right)} \Theta_{\mathbb{P}^m} \to 0$$

vector fields in $\mathbb{P}^1$ have as basis $x_0 \frac{\partial}{\partial x_0}$, $x_0 \frac{\partial}{\partial x_1}$, $x_1 \frac{\partial}{\partial x_0}$. Therefore the holomorphic sections of $\Theta_{\mathbb{F}_n}$ on $U \cap U'$ can be written uniquely as

(6.4) $$\sum_{i \in \mathbb{Z}} z^i \left(a_{i00}\, y_0 \frac{\partial}{\partial y_0} + a_{i01}\, y_0 \frac{\partial}{\partial y_1} + a_{i10}\, y_1 \frac{\partial}{\partial y_0}\right) + \sum_{j \in \mathbb{Z}} b_j z^j \frac{\partial}{\partial z} .$$

where $a_{i00}, a_{i01}, a_{i10}, b_j \in \mathbb{C}$. These sections are holomorphic also on U if only non zero terms occur with $i \in \mathbb{N}$, $j \in \mathbb{N}$. Since

$$\frac{\partial}{\partial y_0'} = \frac{\partial}{\partial y_0}, \quad \frac{\partial}{\partial z'} = -z^2 \frac{\partial}{\partial z} + nzy_0 \frac{\partial}{\partial y_0}, \quad \frac{\partial}{\partial y_1'} = \frac{\partial}{\partial y_1} z^n ,$$

we write a regular section on U' in terms of the first coordinate, and we have

$$\sum_{i \in \mathbb{N}} (z')^i \left(a_{i00}' y_0' \frac{\partial}{\partial y_0'} + a_{i01}' y_0' \frac{\partial}{\partial y_1'} + a_{i10}' y_1' \frac{\partial}{\partial y_0'}\right) + \sum_{j \in \mathbb{N}} b_i'(z')^j \frac{\partial}{\partial z'}$$

$$= \sum_{i \in \mathbb{N}} \left(a_{i00}' y_0 \frac{\partial}{\partial y_0}\right) z^{-i} + \left(a_{i01}' y_0 \frac{\partial}{\partial y_1}\right) z^{n-i} + \left(a_{i10}' y_1 \frac{\partial}{\partial y_0}\right) z^{-n-i} - \sum_{j \in \mathbb{N}} \left(b_j' \frac{\partial}{\partial z}\right) z^{-j+2}$$

$$+ \sum_{j \in \mathbb{N}} \left(nb_j' y_0 \frac{\partial}{\partial y_0}\right) z^{-j+1} .$$

Since $H^1(\Theta_{\mathbb{F}_n}) = H^o(U \cap U', \Theta_{\mathbb{F}_n})/H^o(U, \Theta_{\mathbb{F}_n}) + H^o(U', \Theta_{\mathbb{F}_n})$, we see that

(6.5) $$\left\{ y_1 z^{-1} \frac{\partial}{\partial y_0}, \ldots, y_1 z^{-(n-1)} \frac{\partial}{\partial y_0} \right\} \text{ is a basis of } H^1(\Theta_{\mathbb{F}_n}) .$$

Furthermore, since $H^o(\mathbb{F}_n, \Theta_{\mathbb{F}_n}) = H^o(U, \Theta_{\mathbb{F}_n}) \cap H^o(U', \Theta_{\mathbb{F}_n})$, we have that

(6.6) $$y_0 \frac{\partial}{\partial y_0}, y_0 \frac{\partial}{\partial y_1}, \ldots, z^n y_0 \frac{\partial}{\partial y_1}, \frac{\partial}{\partial z}, z \frac{\partial}{\partial z}, z^2 \frac{\partial}{\partial z} - n y_o z \frac{\partial}{\partial y_o}$$

(and $y_1 (\partial/\partial y_0)$ if $n = 0$) are a basis of $H^o(\mathbb{F}_n, \Theta_{\mathbb{F}_n})$.

Corollary 6.7. $h^1(\Theta_{\mathbb{F}_n}) = n-1$, $h^o(\Theta_{\mathbb{F}_n}) = n+5$ if $n > 0$, 6 if $n = 0$, hence $\mathbb{F}_n \cong \mathbb{F}_m$ if and only if $n = m$. Furthermore, the family defined through the glueing (6.2) is the Kuranishi family of $\mathbb{F}_n$.

Let now $T_k \subset B$ be the determinantal locus

(6.8) $$T_k = \left\{ t \;\middle|\; \operatorname{rank} \begin{pmatrix} t_1 & \cdots & t_{k+1} \\ t_2 & & t_{k+2} \\ \vdots & & \vdots \\ t_{n-k-1} & & t_{n-1} \end{pmatrix} \leq k \right\}$$

We refer to ([Ca 1], §1) for the proof of the following

Proposition 6.9. T_k is an algebraic cone of dimension $\min(2k, n-1)$, and if $t \in T_k - T_{k-1}$, then $X_t \cong \mathbb{F}_{n-2k}$.

Remark 6.10. This example illustrates how the Kuranishi family can be semi-universal only for $t = 0$, and complete for $t \neq 0$. In this case $h^1(\Theta_{X_t})$ has a strict maximum for $t = 0$, and we notice that

(6.11) $h^1(\Theta_{X_t})$ is an uppersemicontinuous function in t, in general, and if $h^1(\Theta_{X_t})$ is constant on B for the Kuranishi family, then the Kuranishi family (cf. 5.2) is also semi-universal for $t \neq 0$.

In this case, as we have seen, $h^o(\Theta_{X_t})$ is not constant: this was, in Wavrik's theorem (5.6), a sufficient condition for the universality of the Kuranishi family. We are going to show that for the above surfaces the Kuranishi family is not universal.

Example 6.12. Take $n = 2$ and our given family $\mathcal{X}$, obtained by the glueing $z' = 1/2$, $y_1' = y_1 z^{-2}$, $y_0' = y_0 + y_1 t_1 z^{-1}$. Consider the local biholomorphism of B to B sending t_1 to $t_1 f(t_1)$, where $f(t_1)$ is a holomorphic function with $f(0) = 1$. Then the pull-back is the family $\mathcal{X}'$ given by the glueing $\zeta' = 1/\zeta$, $\eta_1' = \eta_1 \zeta^{-2}$, $\eta_0' = \eta_0 + \eta_1 t_1 f(t_1) \zeta^{-1}$. But we obtain an isomorphism of $\mathcal{X}'$ with $\mathcal{X}$, compatible with the projections Π, Π' on B, if we set $\zeta = z$, $\zeta' = z'$, $y_0 = \eta_0$, $y_0' = \eta_0'$, $y_1 = \eta_1 f(t_1)$, $y_1' = \eta_1' f(t_1)$. The condition $f(0) = 1$ ensures that the given isomorphism of $\mathbb{F}_n$ with the central fibre X_0 has not been changed.

LECTURE THREE: VARIATIONS ON THE THEME OF DEFORMATIONS

§7. Deformation of automorphisms

As an application of the theorem of Kuranishi, let's assume that G is a finite (or compact) group of biholomorphisms of M. Then we clearly have a natural action of G on $M\times B$, where G acts trivially on the second factor, the base of the Kuranishi family.

If $\sigma \in G$, $t \in B$, clearly σ is holomorphic on X_t if and only if $\sigma_* T_t^{0,1} = T^{0,1}$, i.e. $\Leftrightarrow$ $\sigma_* \varphi(t) = \varphi(t)$. Therefore $G \subset \mathrm{Aut}(X_t)$ for the set $\tilde{B}^G = \{t \mid \sigma_* \varphi(t) = \varphi(t)\}$ (note that it makes sense to talk of σ_* since σ is an automorphism of $M = X_o$). This set is not so weird in general, since $\sigma_* \varphi(t) - \varphi(t)$ is a power series in t, we can only say that it is a complex subspace if G is compact, since, by integration with respect to the invariant measure of G, we can assume $T_M^{1,0}$ to be endowed with a G-invariant Hermitian metric. For $G \ni \sigma$, σ being holomorphic on M, σ_* commutes with $\bar\partial$, but now the G-invariance of the metric implies that σ_* also commutes with $\bar\partial^*$, $\Box$, G, H.

Now G acts naturally on the cohomology groups $H^q(\Theta)$, and we shall write, as customary,

$$H^q(\Theta)^G = \{\eta \mid \in H^q(\Theta) \mid \sigma_* \eta = \eta \ \forall \ \sigma \in G\} .$$

If we identify B with a complex subspace of $H^1(\Theta)$, we have

$$(7.1) \qquad \{t \in B \mid \sigma|_{X_t} \text{ is holomorphic } \forall \ \sigma \in G\} = \tilde{B}^G = B \cap H^1(\Theta)^G = B^G$$

In fact, since $\varphi(t) = t + \frac{1}{2}\ \bar\partial^* G[\varphi(t), \varphi(t)]$,

$$\sigma_* \varphi(t) = \sigma_* t + \frac{1}{2}\ \bar\partial^* G[\sigma_* \varphi(t), \sigma_* \varphi(t)] ,$$

therefore $\sigma_* \varphi(t)$ solves the Kuranishi equation for $\sigma_* t$, and $\sigma_* \varphi(t) = \varphi(\sigma_* t)$. Thus $\sigma_* \varphi(t) = \varphi(t)$ if and only if $t = \sigma_* t$, as we had to show.

But we can be more precise, because we have that

$$B^G = \{t \in H^1(\Theta) \mid \forall \ \sigma \in G\ ,\ t = \sigma_* t,\ H[\varphi(t), \varphi(t)] = 0\}$$

but

$$\sigma_* H[\varphi(t), \varphi(t)] = H[\varphi(\sigma_* t), \varphi(\sigma_* t)] = (\text{if } t = \sigma_* t) = H[\varphi(t), \varphi(t)] ,$$

therefore we get that B^G is a complex subspace of $H^1(\Theta)^G$ defined by $h^2(\Theta)^G$ equations of multiplicity at least 2 and such that their quadratic parts are associated to the symmetric bilinear mapping

$$[\ ,\ ]: H^1(\Theta)^G \times H^1(\Theta)^G \to H^2(\Theta)^G .$$

We get thus a lower bound for $\dim B^G$, and we observe that the family $\mathcal{X}|_{B^G} = \Pi^{-1}(B^G)$ has an action of G which is holomorphic, fibre preserving, and such that the diffeomorphism type of the action is constant.

§8. Deformations of non-degenerate holomorphic maps

Assume we are in the following situation: we are given a family of deformations of M, $\Pi: \mathcal{X} \to B$, and let's assume that, W being a fixed complex manifold, one is given a holomorphic map $F: \mathcal{X} \to W \times B$, such that $\Pi = p_2 \circ F$, $p_2: W \to B$ being the projection on the second factor of the product. This general situation has been considered by Horikawa ([Hor 0], [Hor 1], [Hor 2]), here we shall limit ourselves to the case when

(8.1) $\quad f_t = p_1 \circ F|_{X_t} : X_t \hookrightarrow W$ is generically with injective differential $(f_t)_*$

(here $p_1: W \times B \to W$ is the first projection).

(8.1) is equivalent to saying that $\forall t$ $(f_t)_* \Theta_{X_t} \to (f_t)^* \Theta_W$ is an injective homomorphism of sheaves, where $(f_t)^*$ denotes the analytic pull-back of a coherent sheaf. In particular, for t = 0, we have an exact sequence

$$(8.2) \qquad 0 \to \Theta_{X_o} \xrightarrow{(f_o)_*} (f_o)^* \Theta_W \to N_{f_o} \to 0 \quad .$$

Definition 8.3. The cokernel N_{f_o} of the homomorphism $(f_o)_*$ in (8.2) is called the normal sheaf of the holomorphic map, and $H^o(N_{f_o})$ is called the characteristic system of the map.

Proposition 8.4. There is a linear map $\rho_F: T_{B,o} \to H^o(N_{f_o})$ such that, if ρ is the Kodaira-Spencer map of the given deformation, and ∂ is the coboundary map $\partial: H^o(N_{f_o}) \to H^1(\Theta_{X_o})$ of the long exact cohomology sequence of (8.2), then one has a factorization $\rho = \partial \circ \rho_F$. Such a ρ_F is called the characteristic map of the family.

Proof. Let's take a finite cover (V_α) of $\mathcal{X}$ such that $V_\alpha \cong U_\alpha \times B$ as usual, so that, locally on V_α, if z^α are local coordinates in $U_\alpha \subset X_o$, we can write

$$(8.5) \qquad F(z^\alpha, t) = (f(z^\alpha, t), t) \quad .$$

Let ξ be a tangent vector in $T_{B,o}$ which can be extended as a vector field on B (we always work with (B,o) as a germ of complex space), and let ϑ_α be the lift of ξ to V_α determined by our choice of coordinates (z^α, t) on V_α. We know that $\rho(\xi) = (\vartheta_\alpha - \vartheta_\beta)\big|_{X_o}$, and that in fact (since one can change coordinates), ϑ_α is well defined only up to adding a vertical vector field. The differential F_* sends ϑ_α in a pair (ψ_α, ξ) where ψ_α is a section of $f^*(\Theta_W)$.

Restricting ψ_α to $U_\alpha \times \{0\}$, I get a global section ψ of $N_{f_o} = f_o^*(\Theta_W)/\Theta_{X_o}$ and by definition, since $(f_o)_*(\vartheta_\alpha\big|_{U_\alpha \times \{0\}}) = \psi_\alpha$, we get

$$\partial(\psi) = (\vartheta_\alpha - \vartheta_\beta)\big|_{X_o} = \rho(\xi) .$$

So it suffices to set $\rho_F(\xi) = \psi$, and ρ_F is well-defined and linear. <u>Q.E.D.</u>

The situation considered up to here embodies the classical theory of deformations of plane curves with nodes and cusps, and of surfaces with ordinary singularities, therefore we shall now give a definition which is consistent with the classical one in the second case, but is not a generalization of definition 5.1 (also we shall denote by φ the map $\varphi: M \to W$ corresponding to f_o via the isomorphism $M \cong X$).

<u>Definition 8.6</u>. The characteristic system $H^o(N_\varphi)$ of the map $\varphi: M \to W$ is complete if there exists a smooth deformation of the holomorphic map (i.e. B is smooth) such that $\rho_F: T_{B,o} \to H^o(N_\varphi)$ is surjective.

<u>Remark 8.7</u>. $H^o(N_\varphi)$ is the exact analogue of $H^1(\Theta_M)$ in the case of deformations of a manifold. In general, a sufficient condition in order that the Kuranishi family be smooth is the sharp assumption $H^2(\Theta_M) = 0$: in a similar fashion Horikawa proves the following generalization of a previous theorem of Kodaira:

<u>Theorem 8.8</u>. The characteristic system of a map is complete if $H^1(N_\varphi) = 0$.

In the next paragraph we shall discuss some particular example in the special case when φ is an embedding: before doing so, we just make the following observation (in view of the long exact cohomology sequence associated to (8.2)).

(8.9) a necessary condition in order to deform φ on a complete family $\mathcal{X}$ of deformations of M is that $H^1(\varphi^*(\Theta_W)) \to H^1(N_\varphi)$ be injective.

§9. Examples of embedded deformations and obstructed moduli

Assume now that M is a subvariety of W, that dim W = r, dim M = n, and that M is the locus of zeros of a section of a rank (r-n) vector bundle V. We shall write $V_{|M}$ for $\mathcal{O}_W(V) \otimes \mathcal{O}_M$, and it is an elementary computation to see that, if the $N_{M|W}$ denotes the normal sheaf for the embedding of M into W, then

(9.1) $$N_{M|W} = V_{|M} ,$$

hence we have an exact sequence

$$0 \to \Theta_M \to \Theta_W \otimes \mathcal{O}_M \to V_{|M} \to 0$$

and clearly the characteristic system is complete if there is a surjection

$$H^o(W, V) \twoheadrightarrow H^o(M, V_{|M}) .$$

Example 9.2.

$$W = \mathbb{P}^r , \; V = \bigoplus_{i=1}^{r-n} \mathcal{O}_{\mathbb{P}^r}(m_i) \quad (m_i \geq 2) .$$

The ideal $\mathcal{J}_M$ admits a Koszul resolution

(9.3) $$0 \to \Lambda^{r-n}(V^\vee) \to \Lambda^{r-n-1}(V^\vee) \to \cdots \to \Lambda^2(V^\vee) \to V^\vee \to \mathcal{J}_M \to 0$$

dual to $\Lambda^k V \xrightarrow{\wedge \begin{pmatrix} f_1 \\ \vdots \\ f_{n-r} \end{pmatrix}} \Lambda^{k+1} V$, the f_i's being the sections of $\mathcal{O}_{\mathbb{P}^r}(m_i)$ s.t. $M = \{f_1 = \cdots = f_{n-r} = 0\}$

From (9.3), by induction, follows that

(9.4) $$H^i(\mathcal{J}_M(k)) = 0 \;\; \forall \; k \in \mathbb{Z} \text{ and for } i \leq n .$$

In particular, $H^1(\mathcal{J}_M \cdot V) = 0$, hence, by the sequence

$$0 \to \mathcal{J}_M V \to V \to V_{|M} \to 0 ,$$

we infer that $H^o(\mathbb{P}^r, V)$ goes onto $H^o(M, V_{|M})$. Also, from the sequence

$$0 \to \mathcal{J}_M(k) \to \mathcal{O}_{\mathbb{P}^r}(k) \to \mathcal{O}_M(k) \to 0$$

follows that

(9.5) $$H^i(\mathcal{O}_M(k)) = 0 \quad \forall\, k \in \mathbb{Z} \text{ if } i \le n-1 \quad (\text{e.g. } H^1(V_{|M}) = 0)$$

It will be true that the family given by $H^o(V)$ is <u>complete</u> if and only if, in view of the long exact cohomology sequence associated to (9.1), $H^1(\Theta_{\mathbb{P}^r} \otimes \mathcal{O}_M) = 0$. Now, in view of the Euler sequence (6.3), and of (9.5), we have an exact cohomology sequence giving

(9.6) $$H^1(\Theta_{\mathbb{P}^r} \otimes \mathcal{O}_M) = 0 \quad \text{if } n \ge 3, \text{ or if } n = 2 \text{ and}$$

$$H^2(\mathcal{O}_M) \to H^2(\mathcal{O}_M)^{r+1} \quad \text{is injective.}$$

But in this last case (we use the standard notation Ω^i_M for $\mathcal{O}((\Lambda^i T^{1,0}_M)^\vee)$), by Serre duality an equivalent condition is that the following map be surjective

$$H^o(\Omega^2_M(-1))^{r+1} \to H^o(\Omega^2_M).$$

But, by adjunction, $\Omega^2_M \cong \mathcal{O}_M(\Sigma\, m_i - r - 1)$, and since $H^o(\mathcal{O}_{\mathbb{P}}(k)) \to H^o(\mathcal{O}_M(k))$ is onto $\forall$ k, we get that

(9.7) $$H^1(\Theta_{\mathbb{P}^r} \otimes \mathcal{O}_M) = 0 \quad \text{unless } \Omega^2_M \cong \mathcal{O}_M \quad (\text{i.e. } \sum_1^{r-2} m_i = r+1,$$

an equation which has only the following solutions for the m_i's: (4), (3, 2), (2, 2, 2)). In this last case $H^1(\mathcal{O}_M) = H^o(\Omega^1_M) = 0$ (M is a Kähler manifold via the Fubini-Study metric), but, since $\Omega^2_M \cong \mathcal{O}_S$, $H^2(\Theta_M)$ is dual by Serre duality to

$$H^o((\Omega^1_M) \otimes \Omega^2_M) = H^o(\Omega^1_M) = 0.$$

The conclusion is the following well known fact (cf. [Ser]).

(9.8) If M is a (smooth) complete intersection in $\mathbb{P}^r$ of dimension $n \ge 2$, the characteristic system is complete, and the embedded deformations give a complete deformation, except if $n = 2$ and $\Omega^2_M \cong \mathcal{O}_M$: M is called then a K3 surface, and embedded deformations give a 19-dimensional subvariety of the Kuranishi family, which is smooth of dimension 20.

So, for instance, every small deformation of a smooth surface S of degree m in $\mathbb{P}^3$ is still a surface in $\mathbb{P}^3$: but what happens in the large, according to the following definition?

Definition 9.9. Two manifolds M, M' are said to be a deformation of each other (in the large) if they lie in the same class for the equivalence relation generated by: $M \overset{\mathrm{Def}}{\sim} M' \iff$ there exists a deformation: $\Pi: \mathcal{X} \to B$ of M with irreducible B and such that $\exists\, b \in B$ with X_b isomorphic to M'. If $M \overset{\mathrm{Def}}{\sim} M'$, we shall also say that M' is a direct deformation of M. The upshot ([Hor 3]) is that already for degree $n=5$ the surfaces which are deformations of quintic surfaces need not be surfaces in $\mathbb{P}^3$ any more!

The rough idea is as follows: Horikawa considers all the smooth surfaces such that $p_g = h^o(\Omega^2_S) = 4$, and such that $K^2 = 5$ (K, here and in the following, is a divisor of a (regular here, rational in other cases) section of Ω^2_S): these numerical conditions, as we shall see, are invariant under any deformation, and indeed any complex structure on a surface orientedly homeomorphic to a smooth quintic surface in $\mathbb{P}^3$ must satisfy these conditions.

Studying the behaviour of the rational map $\varphi: S \to \mathbb{P}^3$ associated to the sections of Ω^2_S, Horikawa shows that either

I) φ gives a birational morphism onto a 5-ic or

IIa) φ gives a rational map 2:1 onto a smooth quadric or

IIb) φ is 2:1 to a quadric cone.

Note that for a smooth 5-ic $\Omega^2_S \cong \mathcal{O}_S(1)$, hence φ is just inclusion in $\mathbb{P}^3$ and one is in case I. Surfaces of type I belong to one family of deformations, and their Kuranishi family is smooth of dimension 40: the same holds for surfaces of type IIa). For surfaces of type IIb), instead, $h^1(\Theta_S) = 41$, $h^2(\Theta_S) = 1$ therefore we know that the Kuranishi family gives a hypersurface in $\mathbb{C}^{41}$: Horikawa computes

$$[\ ,\ ]\ H^1(\Theta_S) \times H^1(\Theta_S) \to H^2(\Theta_S),$$

finding that, in suitable coordinates, the associated quadratic polynomial is $z_1 z_2$. Now, by the Morse lemma, there do exist new coordinates on $H^1(\Theta_S)$ s.t. the equation g of B is of the form $g(t) = t_1 t_2 + \psi(t_3, \ldots, t_{41})$, where $\psi(t) = o(t^2)$. Since $H^o(\Theta_S) = 0$, the Kuranishi family is universal and surfaces of type IIb) form, as it is easy to show, a 39-dimensional variety which is contained in the singular locus of B by Horikawa's computation of $[\ ,\ ]$. Now the singular locus of B is given by $t_1 = t_2 = \partial\psi/\partial t_3 = \ldots = \partial\psi/\partial t_{41} = 0$, hence it has dimension 39 iff $\psi \equiv 0$. Thus $g = g_1 g_2$, with $g_i(z) = z_i + o(z)$. The conclusion is now easy: when $g_1 = 0$, $g_2 \neq 0$ we have a surface of type I (a 5-ic), when $g_2 = 0$, $g_1 \neq 0$ we have a surface of type IIa, when $g_1 = g_2 = 0$, we have a surface of type IIb.

We end this section showing a nice example due to Mumford ([Mu 1], cf. also appendix to Chapter V of [Za]) of varieties for which the Kuranishi family is everywhere non reduced. We notice that the terminology most frequently used adopts the following

Definition 9.10. A manifold M is said to have obstructed moduli if $\dim B < h^1(\Theta)$ (if and only if B is not smooth), i.e. if "not all infinitesimal deformations are integrable."

It frequently occurs that B may be singular at 0, but the phenomenon pointed out by Mumford is not so common (at least for the time being).

Example 9.11. ([Mu 1]). Let F be a smooth cubic surface in $\mathbb{P}^3$, E a straight line contained in F (hence $E^2 = -1$, $KE = -1$), and let H be a hyperplane section of F. The linear system $|4H + 2E|$ has no base points (this is clear outside E, on the other hand by the exact sequence $0 \to \mathcal{O}_F(4H) \to \mathcal{O}_F(4H+E) \to \mathcal{O}_E(3) \to 0$ since $H^1(\mathcal{O}_F(4H)) = 0$, we get that $|4H + E|$ has no base points and $H^1(\mathcal{O}_F(4H+E)) = 0$, then we conclude by the exact sequence $0 \to \mathcal{O}_F(4H+E) \to \mathcal{O}_F(4H+2E) \to \mathcal{O}_E(2) \to 0$, where $\mathcal{O}_E(i)$ is the sheaf of degree i on $E \cong \mathbb{P}^1$), so that we can pick a smooth curve C inside $|4H+2E|$. Since the canonical sheaf of C is $\mathcal{O}_C(3H+2E)$, we easily find that

(9.12) $\qquad C \subset F$ is a smooth curve of genus $g = 24$ and degree 14.

Since the normal sheaf of C in F, $N_{C|F}$ is $\mathcal{O}_C(4H+2E)$, which is non special, we get an exact sequence of normal sheaves

(9.13) $$0 \to H^o(N_{C|F}) \to H^o(N_{C|\mathbb{P}^3}) \to H^o(\mathcal{O}_C(3H)) \to 0$$

and also $H^1(N_{C|\mathbb{P}^3}) \cong H^1(\mathcal{O}_C(3H))$, a dual vector space to $H^o(\mathcal{O}_C(2E))$, which has dimension 1 by virtue of the exact sequence

$$0 \to H^o(\mathcal{O}_F(2E)) \to H^o(\mathcal{O}_C(2E)) \to H^1(\mathcal{O}_F(-4H)) = 0 .$$

We see that the hypothesis in theorem 8.8 is not verified, and in fact the characteristic system is not complete, as we shall see. (9.13) gives $\dim_{\mathbb{C}} H^o(N_{C|\mathbb{P}^3}) = 57$, moreover F moves in a 19-dimensional linear system in $\mathbb{P}^3$, C varies in a 37-dimensional linear system on F, hence C belongs to a 56-dimensional family. Mumford shows that C cannot belong to an algebraic family of dimension 57 by the following arguments:

1) if C' is a smooth curve of genus 24, degree 14, $\mathcal{O}_{C'}(4H)$ is non special and has 33 independent sections, so that C' belongs to 2 independent quartic surfaces G, G'. Clearly, C' is not a plane curve, and it is easy to check that C is not contained in any quadric surface.

2) Thus, either

a) C' is not contained in any cubic surface or

b) C' belongs to a (unique) smooth cubic surface or

b′) C' belongs to a singular cubic surface.

3) Assume now that C belongs to an irreducible family of dimension ≥ 57: since condition b′) is a closed condition on the base of the family, there would exist a family of curves C' of type a), with dimension ≥ 57. But in case a), $G \cdot G' = C' + \Gamma$, where Γ is a conic. Thus the complete intersection of G, G', being Cohen-Macauley, is reduced and has at most triple points as singularities, so that G and G' have no singular points in common. Since G and G' intersect transversally at the points of $C' - \Gamma$, we can assume G to be non-singular around C'.

4) By Noether's theorem, not all surfaces of degree 4 contain a conic, hence G belongs to a family of dimension at most 33 (in fact, much less, see [G-H]), moreover it is easily verified that the characteristic system of C' in G has dimension 24, so that the dimension of such pairs $(C' \subset G)$ is at most 57, and since C' belongs to a 1-dimensional system of quartic surfaces, we are done.

Example 9.14 ([Ko 1], [Mu 1]). If M is the blow-up of $\mathbb{P}^3$ with centre a curve C as 9.11, then the base B of the Kuranishi family of deformations is non reduced.

Before even setting up the notations, let's give a useful

Definition 9.15. Let Y be a subvariety of a smooth variety X. We define $\Theta_X(-\log Y)$ to be the sheaf of tangent vectors on X which are tangent to Y (i.e., $\xi \in \Theta_X(-\log Y)$ if and only if, $\mathcal{J}_Y$ being the ideal sheaf of Y, $\forall g \in \mathcal{J}_Y$, $\xi(g) \in \mathcal{J}_Y$).

Remark 9.16. Clearly, $\mathcal{J}_Y \Theta_X \subset \Theta_X(-\log Y)$ and, by the definition, $\Theta_X/\Theta_X(-\log Y)$ is the equisingular normal sheaf $N'_{Y|X}$ of Y in X. ($N'_{Y|X}$ is the usual normal sheaf when Y is smooth, cf. §11.) We have thus the exact sequences

(9.17) $$0 \to \Theta_X(-\log Y) \to \Theta_X \to N'_{Y|X} \to 0 .$$

Moreover the tangent sheaf of Y, Θ_Y is by definition the quotient

$$\Theta_X(-\log Y)/\mathcal{J}_Y\Theta_X .$$

Let's now set up the notation: we have $\Pi : M \to \mathbb{P}^3$ the blow-up map, E is the exceptional divisor $\Pi^{-1}(C)$.

We have therefore two exact sequences

$$0 \to \Theta_{\mathbb{P}^3}(-\log C) \to (\Theta_{\mathbb{P}^3}) \to N_{C/\mathbb{P}^3} \to 0$$

$$0 \to \Theta_M(-\log E) \to \Theta_M \to N_{E|M} \to 0 .$$

First of all, by the exact sequence

$$0 \to \mathcal{J}_C\Theta_{\mathbb{P}^3} \to \Theta_{\mathbb{P}^3}(-\log C) \to \Theta_C \to 0 .$$

since C has genus bigger than 2, $H^0(\Theta_C) = 0$, thus also $H^0(\Theta_{\mathbb{P}^3}(-\log C)) = 0$, since $H^0(\mathcal{J}_C\Theta_{\mathbb{P}^3}) = 0$ (this is the Lie algebra of the group of projectivities leaving C fixed, which is the trivial group since C contains 5 independent points).

Let's look at the exceptional divisor E: E is the projectivized normal bundle of C in $\mathbb{P}^3$, $E = \mathbb{P}(N_{C|\mathbb{P}^3})$, i.e. points in E are lines in the normal bundle of C. In particular, if $\mathcal{O}_E(1)$ is the dual of the tautological invertible sheaf $\mathcal{O}_E(-1) \subset \Pi^*(N_{C|\mathbb{P}^3})$, then $\Pi_*\mathcal{O}_E(1) = N^{\vee}_{C|\mathbb{P}^3}$, ($\vee$ denoting the dual sheaf).

It is easy to verify that

$$N_{E|M} \cong \mathcal{O}_E(-1) , \tag{9.18}$$

hence we have

$$H^0(N_{E|M}) = H^1(N_{E|M}) = 0 . \tag{9.19}$$

<u>Proof</u> of (9.19). Since a sheaf of degree (-1) on $\mathbb{P}^1$ has 0 cohomology in all degree, $\Pi_* N_{E|M} = \mathcal{R}^1\Pi_* N_{E|M} = 0$. <u>Q.E.D.</u>

<u>Corollary 9.20.</u> $H^i(\Theta_M) \cong H^i(\Theta_M(-\log E))$ for $i = 0,1,2$.

<u>Proposition 9.21.</u> $\Pi_*\Theta_M(-\log E) = \Theta_{\mathbb{P}^3}(-\log C)$, $\mathcal{R}^1\Pi_*\Theta_M(-\log E) = 0$, hence

$$H^i(\Theta_M(-\log E)) \cong H^i(\Theta_{\mathbb{P}^3}(-\log E)) \quad \text{for } i = 0,1,2$$

and $H^0(\Theta_M(-\log E)) = 0$.

The proof of proposition 9.21 follows immediately from the following.

Lemma 9.22. Let $\Pi: X \to \mathbb{C}^2$ be the blow up of the origin O, $\mathfrak{m}$ the maximal ideal of the point O, E the exceptional divisor $\Pi^{-1}(0)$. Then

$$\Pi_* \Theta_X(-\log E) = \mathfrak{m}\,\Theta_{\mathbb{C}^2}, \quad R^1\Pi_*\Theta_X(-\log E) = 0 .$$

Proof. X is covered by two affine pieces A, A$'$ with coordinates (u,v), resp. (u$'$,v$'$) and Π is given by $x = u$, $y = uv$ (resp. $x = u'v'$, $y = u'$). Thus

$$\begin{cases} \dfrac{\partial}{\partial u} = \dfrac{\partial}{\partial x} + v\dfrac{\partial}{\partial y} \\[2ex] \dfrac{\partial}{\partial v} = u\dfrac{\partial}{\partial y} \end{cases} \qquad \begin{cases} \dfrac{\partial}{\partial u'} = \dfrac{\partial}{\partial y} + v'\dfrac{\partial}{\partial x} \\[2ex] \dfrac{\partial}{\partial v'} = u'\dfrac{\partial}{\partial x} \end{cases}$$

Since E is defined by the equation $u = 0$ ($u' = 0$), $\Theta(-\log E)$ is generated by

$$u\frac{\partial}{\partial u} = x\frac{\partial}{\partial x} + y\frac{\partial}{\partial y} \quad \text{and by} \quad \frac{\partial}{\partial v} = x\frac{\partial}{\partial y} \quad \text{on A, and by}$$

$$u'\frac{\partial}{\partial u'} = y\frac{\partial}{\partial y} + x\frac{\partial}{\partial x} \quad \text{and} \quad \frac{\partial}{\partial v'} = y\frac{\partial}{\partial x} \quad \text{on A}' .$$

Let ξ be a section of $\Theta_X(-\log E)$ defined in a neighborhood of E: then we can express ξ as

$$\xi = a(u,v)\,u\frac{\partial}{\partial u} + b(u,v)\frac{\partial}{\partial v} \qquad \text{and also}$$

$$\xi = \alpha(u',v')\,u'\frac{\partial}{\partial u'} + \beta(u',v')\frac{\partial}{\partial v'}, \qquad \text{where}$$

$$a(u,v) = \sum_{i,j \geq 0} a_{ij}\, u^i v^j$$

and similarly for b, α, β. We must have

$$ax\frac{\partial}{\partial x} + (ay + bx)\frac{\partial}{\partial y} = (\alpha x + \beta y)\frac{\partial}{\partial x} + (\alpha y)\frac{\partial}{\partial y} .$$

Hence, expressing ξ as a rational section of $\Theta_{\mathbb{C}^2}$, we see that the coefficient of $\partial/\partial x$ can be any function f of the type

$$x \cdot \sum_{i,j \geq 0} a_{ij}\, x^{i-j} y^j ,$$

such that it can also be expressed in the form

$$x \cdot \sum_{h,\ell \geq 0} \alpha_{h\ell}\, x^h \cdot y^{\ell-h} + y \sum_{h,\ell \geq 0} \beta_{h,\ell}\, x^h y^{\ell-h} .$$

It is immediate to see that the coefficient of $\partial/\partial x$ is a power series in x, y, vanishing at the origin, and, by symmetry, the same holds for the coefficient g of $\partial/\partial y$. It is also easy to verify that two such $f, g \in \mathfrak{m}\Theta_{\mathbb{C}^2,0}$ can be chosen arbitrarily. Instead, to show that $R^1\Pi_*(\Theta_X(-\log E)) = 0$, it suffices to show that any section Λ of $\Theta_X(-\log E)$ on $A \cap A'$ can be written as a sum $\Lambda^+ + \Lambda'$, where Λ^+ is regular on A, Λ' is regular on A'. Now, Λ can be written as

$$a(u,v)\, u \frac{\partial}{\partial u} + b(u,v) \frac{\partial}{\partial v},$$

where

$$a(u,v) = \sum_{\substack{i \geq 0 \\ j \in \mathbb{Z}}} a_{ij} u^i v^j$$

can be written as

$$a(u,v) = a^+(u,v) + a'(u,v) \quad a^+ = \sum_{i,j \geq 0} a_{ij} u^i v^i ,$$

and similarly $b(u,v) = b^+(u,v) + b'(u,v)$. Since

$$\Lambda^+ = a^+ u \frac{\partial}{\partial u} + b^+ \frac{\partial}{\partial v}$$

is regular on A, it suffices to verify (we omit this) that Λ' is regular on A'.

Q. E. D.

From 9.20 and 9.21 we infer that

$$H^o(\Theta_M) = 0, \quad H^2(\Theta_M) \cong H^1(N_{C|\mathbb{P}^3}), \quad H^1(\Theta_M) \cong H^o(N_{C|\mathbb{P}^3})/H^o(\Theta_{\mathbb{P}^3}) .$$

These isomorphisms are natural, in fact one can verify that for each deformation of the embedding $C \hookrightarrow \mathbb{P}^3$ the Kodaira-Spencer map for the family of blown up 3-folds is the composition of the characteristic map of the deformation with the surjection of $H^o(N_{C|\mathbb{P}^3}) \twoheadrightarrow H^1(\Theta_M)$ (the kernel $H^o(\Theta_{\mathbb{P}^3})$ is due to the fact that blowing up projectively equivalent curves one obtains isomorphic 3-folds). Now, Kodaira ([Ko 1], thm. 6) proves that every small deformation of M is the blow-up of $\mathbb{P}^3$ with center a curve which is a deformation of C in $\mathbb{P}^3$, thereby showing that the Kuranishi family of M has dimension equal to $56-15 = 41$, whereas, by what we saw, $h^1(\Theta_M) = 42$ for each blow-up M of a curve C as in 9.11. Thus the Kuranishi family B of M is singular at each point ($h^1(\Theta_{M_t})$ being constant for $t \in B$, B is the Kuranishi family for each M_t).

§10. Further variations and further results

We have seen in §9 a 3-dimensional variety such that its Kuranishi family is universal at each point, but its base B is everywhere non reduced. We remarked in §5 that the base B of the Kuranishi family of curves is smooth: for surfaces Kas ([Kas]) found, using Kodaira's theory of elliptic surfaces, an example of a family of elliptic surfaces such that the generic dimension of $H^1(\Theta_{S_t})$ would be strictly bigger than dim B. The family is constructed by deforming a certain class of algebraic surfaces. We suspect that this should not happen for surfaces of general type with $H^1(S, \mathbb{R}) = 0$ and K ample (cf. lecture seven); it is anyhow clarified by Burns and Wahl ([B-W]) how the fact that K_S is not ample, in particular the existence of many curves E such that $\mathcal{O}_E(K_S) \cong \mathcal{O}_E$, forces the dimension of $H^1(\Theta)$ to be bigger than dim B: in particular, using classical results of Segre on the existence of surfaces Σ in $\mathbb{P}^3$ with many nodes (also called conical double points, i.e. with local equation $x^2 + y^2 + z^2 = 0$), they show that the blow-up of Σ at the nodes is a surface S with obstructed deformations (the rough idea being that nodes contribute by 1 to $h^1(\Theta)$, but all the small deformations are still surfaces in $\mathbb{P}^3$). We already noted in the introduction that Zappa ([Zp]) was the first to show that the characteristic system of a submanifold does not need to be complete. His example is as follows (we follow, though, the description of [Mu 3]):

Example 10.1. Let E be an elliptic curve and let V be the rank 2 bundle which occurs as a non trivial extension

$$(10.2) \qquad 0 \to \mathcal{O}_E \to V \to \mathcal{O}_E \to 0$$

(in fact, these extensions are classified by $\mathbb{C}^*$ orbits in $H^1(\mathcal{O}_E) \cong \mathbb{C}$, so there is "only" one non trivial extension). The subbundle $\mathcal{O}_E$ defines a section C of the $\mathbb{P}^1$ bundle $S = \mathbb{P}(V)$ over E, and, since $V/\mathcal{O}_E \cong \mathcal{O}_E$, $N_{C|S} \cong \mathcal{O}_C$. Nevertheless, there is no embedded deformation of C in S, since $H^1(\mathcal{O}_S) \cong H^1(\mathcal{O}_E)$, and, if C′ is algebraically equivalent to C, then there is a divisor L of degree 0 on E such that ($\Pi : S \to E$ being the bundle map) $C' \equiv C + \Pi^*(L)$. But then $\Pi_*(\mathcal{O}_S(C')) = V \otimes \mathcal{O}_E(L)$, and, tensoring 10.2 with $\mathcal{O}_E(L)$, we infer that there are no sections if $L \not\equiv 0$, whereas, for $L \equiv 0$, the condition that the extension splits ensures that 10.2 is not exact on global sections (or geometrically, if

$$h^0(\mathcal{O}_S(C)) \geq 2 , \quad S \cong E \times \mathbb{P}^1 \ !) \quad .$$

Q.E.D.

As far as deformation theory is concerned, the Kodaira-Spencer-Kuranishi results were extended first in the direction of the deformations of isolated singularities (cf. [Po], [Gr 1]), and then the result of Kuranishi was extended to the case of compact complex spaces ([Gr 3], [Dou 3], [Pa]). On the other hand, Grothendieck ([Gro 1], [Gro 2]) contributed significantly to extension of the deformation theory, especially through the construction of the Hilbert schemes, parametrizing projective subschemes with fixed Hilbert polynomials (cf. §19 for a vague idea): his results were extended to the case of (compact subspaces of) complex spaces in Douady's thesis ([Dou 4]). Since the variations in the theme of deformations can be arbitrary, Schlessinger ([Sch]) approached the problem abstractly developing a general theory giving necessary and sufficient conditions for finding "power series solutions", i.e. finding a formal versal deformation space for a deformation functor: this theory is usually coupled with a deep theorem of Artin ([Ar]), giving criteria of convergence for the power series solutions. We don't try to sketch any detail, nor to mention further very interesting work, but we defer the reader to the very interesting article ([Pa]) of Palamodov already quoted in the introduction (and plead guilty for ignoring the post '76 period). We simply remark the importance of Palamodov's theorem 5.6 giving an algebraic description of the higher order terms in the Koranishi equations.

As far as I know, this result has not yet been applied in concrete geometric cases, but its validity should be tested in some example.

LECTURE FOUR: THE CLASSICAL CASE

§11. Deformations of a map and equisingular deformations of the image (infinitesimal theory)

We let, as in §9, $\varphi: X \to W$ be a non-degenerate holomorphic map, and we set $\Sigma = \varphi(X)$. Since W is a smooth variety, we have in general, for every subvariety Σ, the exact sequences

$$(11.1)\qquad \begin{array}{l} 0 \to \Theta_W(-\log \Sigma) \to \Theta_W \to N'_{\Sigma|W} \to 0 \\ 0 \to \Theta_W(-\Sigma) \to \Theta_W(-\log \Sigma) \to \Theta_\Sigma \to 0 . \end{array}$$

On the other hand, by dualizing (i.e., taking $\mathrm{Hom}_{\mathcal{O}_\Sigma}(\,\cdot\,, \mathcal{O}_\Sigma)$) the exact sequence

$$0 \to \check{N}_{\Sigma|W} \to \Omega^1_W \otimes \mathcal{O}_\Sigma \to \Omega^1_\Sigma \to 0$$

where $\check{N}_{\Sigma|W}$ is the conormal sheaf of Σ in W, we get the long exact sequence

$$0 \to \Theta_\Sigma \to \Theta_W \otimes \mathcal{O}_\Sigma \to (\check{N}_{\Sigma|W})^* = N_{\Sigma|W} \to \mathrm{Ext}^1(\Omega^1_\Sigma, \mathcal{O}_\Sigma) \to 0 .$$

which splits into the short exact sequences

$$(11.2)\qquad \begin{array}{l} 0 \to \Theta_\Sigma \to \Theta_W \otimes \mathcal{O}_\Sigma \to N'_{\Sigma|W} \to 0 \\ 0 \to N'_{\Sigma|W} \to N_{\Sigma|W} \to \mathrm{Ext}^1(\Omega^1_\Sigma, \mathcal{O}_\Sigma) \to 0 . \end{array}$$

Example 11.3. Assume Σ is a hypersurface in W, locally defined by the equation $f(x_1, \ldots, x_n) = 0$. Then $N_{\Sigma|W} \cong \mathcal{O}_\Sigma(\Sigma)$, and N'_Σ is the subsheaf defined by the ideal sheaf $(\partial f/\partial x_1, \ldots, \partial f/\partial x_n)$. Thus if g is a section of $N'_{\Sigma|W}$, then $f_t = f(x) + tg(x) = 0$ gives an infinitesimal deformation of Σ which is "equisingular," i.e., modulo (t^2), the locus of zero has not changed. In fact, if

$$g(x) = \sum_i \frac{\partial f}{\partial x_i} \cdot u_i(x) \quad ,$$

then, setting $u(x) = (u_1(x), \ldots, u_n(x))$, we have that $f_t(x) \equiv f(x + tu(x)) \pmod{t^2}$ by Taylor expansion.

Definition 11.4. The morphism φ is said to be stable if the direct image sheaf $\varphi_*(N_\varphi)$ is isomorphic to the equisingular sheaf $N'_{\Sigma|W}$.

Remark 11.5. By looking at the stalks of $N'_{\Sigma|W}$ and of $\varphi_*(N_\varphi)$ at p, a smooth point of Σ, such that φ has differential of maximal rank at the points in $\varphi^{-1}(p)$, we see immediately that if φ is stable, then φ is birational onto its image (otherwise one should equip the locus Σ with a scheme structure with nilpotent elements). We shall assume from now on φ to be birational onto its image.

Proposition 11.6. Assume $\dim X = 1$, and that p is a singular point of Σ: then if φ is stable, then p is an ordinary double point (node) and $\dim W = 2$.

Proof. In fact, the rank of $N'_{\Sigma|W}$ at p ($\operatorname{rank}_p \mathfrak{F} = \dim_{\mathbb{C}} \mathfrak{F}/\mathfrak{m}_p\mathfrak{F}$, $\mathfrak{m}_p$ being the maximal ideal of p) is $n = \dim W$, since $\partial/\partial x_1, \partial/\partial x_2, \ldots, \partial/\partial x_n$ generate $N'_{\Sigma|W}$ locally, while vector fields in Θ_Σ vanish at p (cf. [Ro], thm. 3.2); whereas the rank of $\varphi_*(N_\varphi)$ at p is just the sum

$$\sum_{\varphi(q) = p} \dim_{\mathbb{C}} \frac{(N_\varphi)_q}{\varphi^* \mathfrak{m}_p (N_\varphi)_q}$$

If $\varphi(q) = p$, then we can take local holomorphic coordinates t at q, $(x_1, \ldots, x_n)$ at p, such that

$$\varphi(t) = \left(t^{a_1}, t^{a_1 + a_2} + \ldots\ldots\ldots, t^{a_1 + a_2 + \ldots + a_r} + \ldots\ldots, 0, 0 \right)$$

where $a_i > 0$, and r is the smallest dimension of a local smooth subvariety containing the branch of Σ corresponding to q (clearly, $a_1 = 1 \Leftrightarrow r = 1$, and $r \geq a_1$). It is easy then to seé that

$$\frac{\dim (N_\varphi)_q}{\varphi^* \mathfrak{m}_p (N_\varphi)_q} \geq n a_1 - 1 \ .$$

Hence if this dimension has to be less than n, we get already $a_1 = 1$; moreover, since the sum over all these points q has to be less than n, we infer that $n = 2$ and there are exactly two smooth branches. Either the two branches are transversal, and we have a node ($x_1 x_2 = 0$ in suitable coordinates on W), or we have a double point of type (we set $x = x_1$, $y = x_2$) $y^2 = x^{2k+2}$ ($k \geq 1$). In this case

$$\frac{\varphi_*(N_\varphi)}{\mathfrak{m}_p^2 \varphi_*(N_\varphi)} ,$$

though, has dimension 4 whereas

$$\frac{N_{\Sigma,p}}{\mathfrak{m}_p^2 N_{\Sigma,p}} = \frac{(y, x^{2k+1})}{(y, x^{2k+1})(y^2, xy, x^2) + (y^2 - x^{2k+2})}$$

has dimension 5 (since $y, xy, y^2, x^{2k+1}, x^{2k+1} \cdot y$ are a $\mathbb{C}$-basis). Q. E. D.

Example 11.7. The morphism $t \mapsto (t^2, t^3)$, giving an ordinary cusp, is not stable since in fact the deformation $t \mapsto (a_0 + t^2, b_0 + b_1 t + t^3)$ gives a node: in fact $\partial x_1/\partial t = 2t$, $\partial x_2/\partial t = b_1 + t^2$, and $t = \pm\sqrt{-b_1}$ are two points of X mapping to the same point.

Remark 11.8. We refer to [Math 1, 2] for a thorough and general discussion about stable map germs: here we shall limit ourselves in the sequel to discuss ordinary singularities when $\dim X = 2$. Before dealing with this special case, let's see what is true in general.

Theorem 11.9. There is a natural injective homomorphism of $N'_{\Sigma|W}$ to $\varphi_* N_\varphi$, if you assume φ to be finite and birational onto Σ.

Proof. We have the two following exact sequences, with the homomorphism ψ induced by pull-back $\varphi^*: \Theta_\Sigma \to \varphi_* \Theta_X$.

(11.10)
$$\begin{array}{ccccccccc} 0 \to \Theta_\Sigma & \to & \Theta_W \otimes \mathcal{O}_\Sigma & \to & N'_{\Sigma|W} & \to 0 & & \\ & & & & \downarrow \psi & & & \\ 0 \to \varphi_*\Theta_X & \to & \Theta_W \otimes \varphi_* \mathcal{O}_W & \to & \varphi_* N_\varphi & \to & R^1 \varphi_* \Theta_X & = 0 \end{array}$$

and we have to verify that $\psi(\Theta_\Sigma) \subset \varphi_* \Theta_X$. I.e., this is what we need to verify: if $\mathcal{J}_\Sigma$ is the ideal sheaf of Σ, and $(x_1, \ldots, x_n)$ are coordinates in W, whenever $a_1(x), \ldots, a_n(x)$ are functions such that $\forall f \in \mathcal{J}_\Sigma$,

$$\sum_{i=1}^{n} a_i(x) \frac{\partial f}{\partial x_i} \in \mathcal{J}_\Sigma ,$$

then there do exist, for each point y s.t. $\varphi(y) = x$, functions $\beta_1(y), \ldots, \beta_m(y)$ ($m = \dim X$, $y = (y_1, \ldots, y_m)$ are coordinates on X) such that

$$a_i(x) = \sum_{j=1}^{m} \beta_j(y) \left(\frac{\partial x_i}{\partial y_j}\right) .$$

Since X is smooth, by Hartog's theorem it will suffice to show the existence of such functions outside a subvariety Γ of codimension at least 2 in X.

We first remark that, since we are assuming φ to be birational onto Σ, the subvariety $Z \subset X$ where φ is not of maximal rank has image $\varphi(Z) \subset \mathrm{Sing}(\Sigma)$ (if $\varphi: \mathbb{C}^n_{,0} \to \mathbb{C}^n_{,0}$ has local degree 1, then it is a local biholomorphism). If

$x \in \Sigma - \mathrm{Sing}(\Sigma)$, there is nothing to prove, otherwise there exists a codimension 2 subvariety Δ of Σ with

i) $\Delta \subset \mathrm{Sing}(\Sigma)$,

ii) if $x \in \mathrm{Sing}(\Sigma) - \Delta$, Σ is locally biholomorphic to $\Sigma^1 \times \mathbb{C}^{m-1}$ where Σ^1 is a curve in $\mathbb{C}^{n-m+1}$,

iii) $\varphi^{-1}(\Delta) = \Gamma$ has codimension at least 2 in X (this follows from the assumption that φ be finite,

iv) if $x \in \mathrm{Sing}(\Sigma) - \Delta$, $y \in \varphi^{-1}(x)$, then there are coordinates $(y_1, \dots, y_m)$ around y such that, using the local biholomorphism $\Sigma \cong \Sigma^1 \times \mathbb{C}^{m-1}$, $\varphi(y) = (\varphi^1(y), y_2, \dots, y_m)$.

By our previous remark and iv) above, it suffices to prove our result in the case when $\dim X = \dim \Sigma = 1$. In this case, we denote by t a local coordinate at a point y of X, according to tradition, and we may assume that, $\forall\ 1 \le i < j \le n$, projection on the (i,j) coordinates maps Σ birationally to a plane curve of equation $F_{ij}(x_i, x_j) = 0$. Since $F_{ij} \in \mathcal{J}_\Sigma$, we have

$$a_i(x(t)) \frac{\partial F}{\partial x_i}(x_i(t), x_j(t)) + a_j(x(t)) \frac{\partial F}{\partial x_j}(x_i(t), x_j(t)) \equiv 0 .$$

Assume we show that there exists $\forall\ i, j$, a meromorphic function $v(t)$ with

$$a_i(x(t)) = v(t) \frac{dx_i(t)}{dt}, \qquad a_j(x(t)) = v(t) \frac{dx_j(t)}{dt}$$

then all the (2×2) minors of the matrix

$$\begin{pmatrix} \dfrac{dx_1(t)}{dt} & \cdots & \dfrac{dx_n(t)}{dt} \\ \\ a_1(x(t)) & \cdots & a_n(x(t)) \end{pmatrix}$$

vanish, and we can conclude that there is a $v(t)$ with $a_i(x(t)) = v(t)(dx_i(t)/dt)$ for each i, provided the following holds true.

<u>Lemma 1.12.</u> Let Σ be a germ of plane curve singularity, with equation $f(x, y) = 0$ and let $X = \varphi_1(t)$, $y = \varphi_2(t)$ be a parametrization of a branch of Σ. Then, if $a_1(t)$, $a_2(t)$ are functions such that

$$a_1(t) \frac{\partial f}{\partial x}(\varphi_1(t), \varphi_2(t)) + a_2(t) \frac{\partial f}{\partial y}(\varphi_1(t), \varphi_2(t)) \equiv 0 ,$$

there does exist a meromorphic function $v(t)$ with

$$a_i(t) = v(t)\,\frac{d\varphi_i(t)}{dt} \qquad (i = 1, 2)\ .$$

Proof. Write $f = f_1 f_2$ where $f_1 = 0$ is the local equation of the given branch. Clearly

$$\frac{\partial f}{\partial x}(\varphi_1(t), \varphi_2(t)) = \frac{\partial f_1}{\partial x}(\varphi_1(t), \varphi_2(t)) \cdot f_2(\varphi_1(t), \varphi_2(t))$$

and analogously for $\partial f/\partial y$: since $f_2(\varphi_1(t), \varphi_2(t) \not\equiv 0$, we can indeed assume Σ to have only one branch. Without loss of generality we may assume

$$\varphi_1(t) = t^m\ , \quad \varphi_2(t) = g(t) = t^{m+c} + \cdots$$

(m is the multiplicity, c the local class). As classical, we use a base change $\varphi: \mathbb{C}^2 \to \mathbb{C}^2$ sending (t,y) to $(x = t^m, y)$: then $\varphi^{-1}(\Sigma)$ consists of m smooth branches, of equation $y - g(t\epsilon^i) = 0$, with $\epsilon = \exp(2\pi\sqrt{-1}/m)$, $i = 1,\dots,m$ (the first of these branches coincides with the given parametrization). Clearly the pull-backs $\varphi^*(\partial f/\partial y)$ and $\varphi^*(\partial f/\partial x)$ coincide, respectively, with

$$\frac{\partial f(t,y)}{\partial y} \quad \text{and} \quad \frac{1}{m t^{m-1}}\,\frac{\partial f(t,y)}{\partial t}$$

Since

$$f(t,y) = \prod_{i=1}^{m} (y - g(t\epsilon^i))\ ,$$

by assumption

$$a_1(t)\,\frac{\partial f(t,y)}{\partial t} + a_2(t)\, m t^{m-1}\,\frac{\partial f(t,y)}{\partial y}$$

vanishes identically after plugging in $y = g(t)$. We get

$$-a_1(t) \sum_{i=1}^{m} \prod_{j\neq i} (y - g(t\epsilon^j))\,\epsilon^i \frac{\partial g}{\partial t}(t\epsilon^i) + a_2(t)\, m t^{m-1} \sum_{i=1}^{m} \prod_{j\neq i} (y - g(t\epsilon^j))$$

and plugging in $y = g(t)$, we obtain

$$-a_1(t)\,\frac{\partial g(t)}{\partial t} + a_2(t)\, m t^{m-1} \equiv 0\ .$$

<u>Q. E. D. for the lemma.</u>

Now

$$a_i(x(t)) = v(t)\,\frac{dx_i(t)}{dt}\ ;$$

let m be the multiplicity of the branch (i.e. $m = \min \operatorname{ord}_t x_i(t)$)) and assume $x_1(t) = t^m$: then $\operatorname{ord}_t a_i(x(t)) \geq m$, hence $\operatorname{ord}_t v(t) \geq 0$ and g is holomorphic. It remains to be proved that the given homomorphism of $N_{\Sigma|W}$ into $\varphi_* N_\varphi$ is injective. In view of (11.10) we have to show that if a section ξ of $\Theta_W \otimes \mathcal{O}_\Sigma$ lies in the image of $\varphi_* \Theta_X$, then its image in $N_{\Sigma|W}$ equals zero.

By 11.2 it suffices to show that its image in $N_{\Sigma|W} = \mathrm{Hom}(\check{N}_{\Sigma|W}, \mathcal{O}_\Sigma)$ is zero. Let ν be a section of $\check{N}_{\Sigma|W}$: since ξ is tangent to Σ at the smooth points of Σ, $\langle \xi, \nu \rangle$ vanishes on an open dense set, thus $\langle \xi, \nu \rangle \equiv 0$, Σ being reduced. Q.E.D.

We shall not pursue here the analogue of Kuranishi's theory for these deformation theories (cf. [Sch], [Wah], [B-W], [Pa]), in fact, as we have shown already and will see in the sequel, it is very hard to compute the obstructions in almost all the examples, whereas geometry can help to find a complete family of deformations.

§12. Surfaces with ordinary singularities

Here X is a smooth surface and will hence be denoted by S, $\varphi : S \to \Sigma \subset W$, where W is a smooth 3-fold is a finite map, birational onto its image Σ, which possesses only the following type of singularities:

i) nodal curve ($xy = 0$ in local holomorphic coordinates)

ii) triple points ($xyz = 0$ in local coordinates)

iii) pinch points ($x^2 - zy^2 = 0$ in local coordinates).

Δ will be the double curve ($=$ Sing(Σ)) of Σ, smooth at points of type i), iii), with local equations $x = y = 0$, and with a triple transversal point at each triple point. We let $D = \varphi^{-1}(\Delta) \subset S$, and notice that a pinch point p' has just one inverse image point p, where we can choose local coordinates (u,v) such that

$$(12.1) \qquad \varphi(u,v) = (uv, v, u^2)$$

hence in particular $D = \{(u,v) \mid v = 0\}$.

<u>Proposition 12.2.</u> If Σ has ordinary singularities, the morphism φ is stable (i.e., ch. 11.4, $\varphi_*(N_\varphi) \cong N'_{\Sigma|W}$).

<u>Proof.</u> In view of 11.6 and 11.9, it suffices to consider the case of triple and pinch points, and to prove that $\mathcal{N}'_{\Sigma|W}$ goes onto $\varphi_* N_\varphi$. To do this, we shall explicitly compute these two sheaves.

<u>Lemma 12.3.</u> $N'_{\Sigma|W} \subset N_{\Sigma|W} \cong \mathcal{O}_\Sigma(\Sigma)$ is the subsheaf of sections g vanishing on Δ and satisfying the further linear condition: $\partial g/\partial y = 0$ at the pinch points.

Proof. $g \in N'_{\Sigma|W}$ iff it belongs locally to the Jacobian ideal of Σ, i.e. (x,y) for the nodal points, (xy, yz, xz) for the triple points, (x, y^2, yz) for the pinch points. At the pinch points, $g \in (x, y^2, yz) \Leftrightarrow g = xg_1 + yg_2$ with $g_2(0,0,0) = 0 \Leftrightarrow$ $\partial g/\partial y\,(0,0,0) = 0$. Q.E.D.

Remark 12.4. Since g vanishes on Δ, clearly $\partial g/\partial z = 0$ at a pinch point p'. Hence the condition $\partial g/\partial y = 0$ can be formulated also as: $\xi(g) = 0$ for each tangent vector at p' lying in the tangent cone to Σ, $\{x^2 = 0\}$. Clearly this last formulation is independent of the choice of coordinates.

Lemma 12.5. Let $p_1,\ldots,p_k$ be the points of S mapping to the pinch points $p'_1,\ldots,p'_k$ of Σ: then $N_\varphi \cong \prod_{i=1}^k \mathfrak{m}_{p_i}\, \mathcal{O}_S(\varphi^*\Sigma - D)$.

Proof. By 11.9 and 12.3 we know that N_φ coincides with $\varphi^*(N'_{\Sigma|W})$ except at a finite number of points, and that $\varphi^*(N'_{\Sigma|W})$ equals to $\mathcal{J}'\mathcal{O}_S(\varphi^*\Sigma - D)$, where $\mathcal{J}'$ is an ideal sheaf of a 0-dimensional scheme. Hence N_φ is also of the form $N_\varphi = \mathcal{J}\mathcal{O}_S(\varphi^*\Sigma - D)$, with $\dim(\mathrm{supp}(\mathcal{O}_{S/\mathcal{J}})) = 0$. To determine the ideal $\mathcal{J}$, we first notice that $\mathrm{supp}(\mathcal{O}_{S/\mathcal{J}}) = \{p_1,\ldots,p_k\}$, then that, at p_i, $N_\varphi \cong \mathrm{coker}\, \mathcal{O}^2 \xrightarrow{\Phi} \mathcal{O}^3$ where Φ = differential of φ, sends a pair (g_1, g_2) to a triple $f_1 = vg_1 + ug_2$, $f_2 = g_2$, $f_3 = 2ug_1$. The homomorphism of $\mathcal{O}^3 \to \mathfrak{m}_{p_i}$ sending (f_1, f_2, f_3) to $(2uf_1 - vf_3)$ clearly gives an isomorphism of N_φ with $\mathfrak{m}_{p_i}$. Q.E.D.

We can now finish the proof that $\varphi_*(N_\varphi) = N'_{\Sigma|W}$: in fact $\varphi_*\mathcal{O}_S(-D) = \mathcal{O}_\Sigma(-\Delta)$, as it is easy to see, whereas at the pinch points $\mathfrak{m}_{p_i}\mathcal{O}_S(-D) = (uv, v^2)$, whereas $g \in N'_{\Sigma|W}$ iff $g = xg_1 + yg_2$ with $g_2 \in \mathfrak{m}_{p'_i}$; as we have already seen, $\varphi^*(g) = uv\,\varphi^*(g_1) + v\,\varphi^* g_2 \in \mathfrak{m}_{p_i}\mathcal{O}_S(-D)$, and we can conclude since both sheaves $\varphi_*(N_\varphi) \supset N_{\Sigma|W}$ have codimension 1 in $\mathcal{O}_\Sigma(\Sigma - \Delta)$.

Q.E.D. for Proposition 12.2.

Corollary 12.6. If Σ has ordinary singularities only in the smooth 3-fold W, then there exists an exact sequence

$$0 \to H^0(\mathcal{O}_S) \to H^0(\varphi^*\mathcal{O}_W) \to H^0(N'_{\Sigma|W}) \to H^1(\mathcal{O}_S) \to H^1(\varphi^*\mathcal{O}_W)$$

$$\to H^1(N_{\Sigma|W}) \to H^2(\mathcal{O}_S) \to H^2(\varphi^*\mathcal{O}_W) \to H^2(N'_{\Sigma|W}) \to 0$$

Proof. Obvious from the Leray spectral sequence for the finite map φ.

Q.E.D.

Definition 12.7. $\Theta_W(\Sigma)(-\Delta - c')$, where c' stands for "cuspidal conditions," is defined to be the inverse image of $N'_{\Sigma|W}$ under the surjective homomorphism $\Theta_W(\Sigma) \to \Theta_\Sigma(\Sigma) = N_\Sigma$.

The heuristic explanation for $\Theta_W(\Sigma)(-\Delta - c')$ (cf. [Ko 2]) is as follows: assume that you deform the singular locus of Σ by deforming with a parameter t the local coordinates x, y, z; then if $X(t) = x + t\xi + \cdots$, $Y(t) = y + t\eta + \ldots$, $Z(t) = z + t\zeta + \ldots$, the local equation of Σ changes as follows:

$$\begin{cases} XY = xy + t(\xi y + \eta x) + \cdots \\ XYZ = xyz + t(\xi yz + \eta xz + \zeta xy) + \cdots \\ X^2 - Y^2 Z = x^2 - y^2 z + t(2\xi x - 2zy\eta - \zeta y^2) + \cdots \end{cases}$$

Hence, if $f = 0$ was the old equation, the new one is of the form $f + tg + \ldots$, where g is a section of $\Theta_W(\Sigma)$ vanishing on Δ and satisfying the cuspidal conditions.

We clearly have an exact sequence (f is a section with $\operatorname{div}(f) = \Sigma$).

$$(12.8) \qquad 0 \to \Theta_W \xrightarrow{\ f\ } \Theta_W(\Sigma)(-\Delta - c') \to N'_{\Sigma|W} \to 0,$$

and Kodaira, after Severi, gives the following (cf. [Ko 2]).

Definition 12.9. Σ is said to be regular if $H^1(\Theta_W(\Sigma)(-\Delta - c')) = 0$, and semi-regular if $H^1(\Theta_W(\Sigma)(-\Delta - c')) \to H^1(N_{\Sigma|W})$ is the zero map.

Remark 12.10. The two definitions coincide if $H^1(\Theta_W) = 0$, e.g. for $W = \mathbb{P}^3$.

We have the following.

Theorem 12.11 (Kodaira, [Ko 2]). If Σ is semi-regular the characteristic system of the map $\varphi: S \to \Sigma$ is complete; moreover, there is a smooth semi-universal family $\{\varphi_t\}$ of deformations of $\varphi: S \to \Sigma$ such that the characteristic system is complete also for $t \neq 0$.

Unfortunately, the condition of semi-regularity is a very strong assumption upon $\Sigma \subset W$: we shall, following Kodaira ([Ko 2], [Ko 3]), consider from now on only the classical case where $W = \mathbb{P}^3$, and regularity coincides with semi-regularity.

Theorem 12.12. If Σ is a surface in $\mathbb{P}^3$ of degree n with ordinary singularities, Σ is (semi)-regular if and only if the cuspidal conditions are independent on the space of polynomials of degree n vanishing on the double curve Δ of Σ (i.e.,

$$0 \to H^0(\mathcal{O}_{\mathbb{P}^3}(nH)(-\Delta - c')) \to H^0(\mathcal{O}_{\mathbb{P}^3}(nH)(-\Delta) \xrightarrow{\partial g/\partial y_i} \bigoplus_i \mathbb{C}_{p_i'} \to 0$$

is exact, where H is the hyperplane divisor on $\mathbb{P}^3$).

Proof. By assumption

$$H^1(\mathcal{O}_{\mathbb{P}^3}(nH)(-\Delta - c') = H^1(\mathcal{O}_{\mathbb{P}^3}(nH)(-\Delta)) \quad .$$

By the exact sequence

$$0 \to \mathcal{O}_{\mathbb{P}^3} \to \mathcal{O}_{\mathbb{P}^3}(nH)(-\Delta) \to \mathcal{O}_{\Sigma}(nH)-\Delta) \to 0$$

we have thus to show that $H^1(\mathcal{O}_{\Sigma}(nH-\Delta)) = 0$. Denoting still by H the pull-back of a hyperplane, we have $H^1(\mathcal{O}_{\Sigma}(nH-\Delta)) = H^1(\mathcal{O}_S(nH-D))$. Since, by adjunction, the canonical divisor on S is $(n-4)H - D$, by Serre duality, our space is dual to $H^1(\mathcal{O}_S(-4H))$, which is zero since H is ample (e.g. by Kodaira's vanishing theorem).

Q. E. D.

The preceding criterion of regularity is not so easy to apply directly, thus the usual method is to relate the equisingular deformations of Σ to the (equisingular) deformations of Δ (observing that sections of $\mathcal{O}_{\mathbb{P}^3}(nH)$ vanishing on Δ of order 2 give trivial infinitesimal deformations of Δ).

We have the usual exact sequences (cf. 9.17)

(12.13.i) $$0 \to \Theta_{\mathbb{P}^3}(-\log \Delta) \to \Theta_{\mathbb{P}^3} \to N'_{\Delta|\mathbb{P}^3} \to 0$$

(12.13.ii) $$0 \to \Theta_{\Delta} \to \Theta_{\mathbb{P}^3} \otimes \mathcal{O}_{\Delta} \to N'_{\Delta|\mathbb{P}^3} \to 0$$

and moreover ([Ko 3], thm. 4).

Theorem 12.14. There exists an exact sequence

$$0 \to \mathcal{O}_{\mathbb{P}^3}(nH)(-2\Delta) \to \mathcal{O}_{\mathbb{P}^3}(nH)(-\Delta - c') \to N'_{\Delta|\mathbb{P}^3} \to 0 \ .$$

Idea of Proof (see loc. cit. for details). Let $\tilde{\Delta}$ be Δ - {triple points and pinch points of Σ}, and let $\tilde{N}$ be the normal bundle of $\tilde{\Delta}$ in $\mathbb{P}^3$. Since the conormal sheaf $N^\vee$ of Δ is just $\mathcal{O}_{\mathbb{P}^3}(-\Delta)/\mathcal{O}_{\mathbb{P}^3}(-2\Delta)$, the basic claim is that there exists an isomorphism of $\tilde{N}$ into $N^\vee|_{\tilde{\Delta}} \otimes \mathcal{O}_{\mathbb{P}^3}(nH)$: and after that one has to check that this isomorphism extends at the cuspidal points onto the subsheaf defined by the cuspidal conditions, and at the triple points there is a similar verification. Since $\tilde{N}$ and $N^{\vee}|_{\tilde{\Delta}}$ are dual bundles, the key point is that the equation f of Σ, locally of the form $xy = 0$, induces locally two sections of $N^{\vee}|_{\tilde{\Delta}}$, and globally a non vanishing section of $\Lambda^2 N^{\vee}|_{\tilde{\Delta}} \otimes \mathcal{O}_{\mathbb{P}^3}(nH)$, thereby inducing a non degenerate pairing $\tilde{N} \times \tilde{N} \to \mathcal{O}_{\tilde{\Delta}}(nH)$, hence the desired isomorphism. <u>Q. E. D.</u>

The important feature of 12.14 is that the left term of the exact sequence depends only upon the double curve Δ and the degree n of Σ, but not upon Σ. Moreover, given any curve Δ in $\mathbb{P}^3$, by Serre's theorem ([Se]), there is an integer

(12.15) $\qquad n_o(\Delta) = \min\{n \mid H^i(\mathcal{O}_{\mathbb{P}^3}(kH)(-2\Delta)) = 0 \ \ \forall\ i = 1, 2,\ k \geq n\}$.

<u>Theorem 12.16.</u> Let Σ be a surface of degree n with ordinary singularities in $\mathbb{P}^3$ having Δ as double curve; if $n \geq n_o(\Delta)$, Σ is regular if and only if $H^1(N'_{\Delta|\mathbb{P}^3}) = 0$. In particular Σ is regular if $H^1(\Theta_{\mathbb{P}^3} \otimes \mathcal{O}_{\Delta}) = 0$.

<u>Proof.</u> By the cohomology sequence attached to 12.14, and by 12.15,

$$H^1(\Theta_{\mathbb{P}^3}(nH)(-\Delta - c')) \cong H^1(N'_{\Delta|\mathbb{P}^3}) .$$

The other assertion follows from (12.13.ii). <u>Q. E. D.</u>

<u>Theorem 12.17.</u> Let Σ be a surface of degree n with ordinary singularities in $\mathbb{P}^3$ having Δ as double curve, and let $\hat{\Delta}$ be the normalization of Δ.

i) If T is the divisor on $\hat{\Delta}$ given by the sum of the triple points, and H is the hyperplane divisor, then if $n \geq n_o(\Delta)$ and $\mathcal{O}_{\hat{\Delta}}(H-T)$ is non-special on (every component of) $\hat{\Delta}$, then Σ is regular.

ii) If there exists a surface Σ' of degree n' containing Δ, and such that the divisor $\varphi^*(\Sigma')$ on S has no multiple components, then $n_o(\Delta) \leq n + n' - 3$.

<u>Proof.</u> i) by 12.16 it suffices to show $H^1(\Theta_{\mathbb{P}^3} \otimes \mathcal{O}_{\Delta}) = 0$. By the Euler sequence (6.3) tensored with $\mathcal{O}_{\Delta}$, it suffices to show that $H^1(\mathcal{O}_{\Delta}(1)) = 0$. Now, if $\psi : \hat{\Delta} \to \Delta$ is the normalization map, $\psi_*(\mathcal{O}_{\hat{\Delta}}(H-T)) = \mathcal{m}_T \mathcal{O}_{\Delta}(1)$, where $\mathcal{m}_T$ is the ideal sheaf of the triple points. Hence $H^1(\mathcal{m}_T \mathcal{O}_{\Delta}(1)) = 0$ and we are done by the exact sequence

$$0 \to \mathcal{m}_T \mathcal{O}_{\Delta}(1) \to \mathcal{O}_{\Delta}(1) \to \tau \to 0$$

where τ is a skyscraper sheaf with stalk $\cong \mathbb{C}$ at each triple point.

ii) let k be an integer $\geq n+n'-3$ and consider the exact sequence

$$0 \to \mathcal{O}_{\mathbb{P}^3}(kH-\Sigma) \to \mathcal{O}_{\mathbb{P}^3}(kH)(-2\Delta) \to \mathcal{O}_\Sigma(kH-2\Delta) \to 0$$

Since $H^i(\mathcal{O}_{\mathbb{P}^3}(kH-\Sigma)) = H^i(\mathcal{O}_{\mathbb{P}^3}((k-n)H)) = 0$ for $i=1,2$, it suffices to show the vanishing of $H^1(\mathcal{O}_\Sigma(kH-2\Delta))$. Since $\mathcal{O}_\Sigma(-2\Delta) = \varphi_* \mathcal{O}_S(-2D)$, we want the vanishing of $H^1(\mathcal{O}_S(kH-2D))$. As in 12.12, since $K_S = (n-4)H - D$, the dual vector space is, by Serre duality, $H^1(\mathcal{O}_S(-(k-n+4)H+D))$. By assumption, $n'H \equiv \varphi^*(\Sigma') = D+\Gamma$, hence we want the vanishing of $H^1(\mathcal{O}_S(-aH-\Gamma))$, where $a = k-n+4-n' \geq 1$. But $|aH+\Gamma|$ maps to a surface and $|aH+\Gamma|$ contains a reduced connected divisor, hence one can apply the Ramanujam vanishing theorem (cf. e.g. [Bo], [Ram]).

<u>Q. E. D.</u>

By 12.14, if $n \geq n_o(\Delta)$, then $H^o(\mathcal{O}_{\mathbb{P}^3}(nH)(-\Delta-c'))$ goes onto $H^o(N'_{\Delta|\mathbb{P}^3})$: on the other hand, by (12.8) this surjective homomorphism factors through the one onto $H^o(N'_{\Sigma|\mathbb{P}^3})$, which has the subspace $\mathbb{C}f$ as its kernel ($f=0$ being the equation of Σ). Assume now Δ to be smooth (thus Σ has no triple points): then if the characteristic system of Σ is complete, and $n \geq n_o(\Delta)$, then also the characteristic system of Δ is complete; moreover, Kodaira (loc. cit., p. 246) proves the converse.

<u>Theorem 12.18</u>. Let Σ be a surface of degree n in $\mathbb{P}^3$ with ordinary singularities and smooth double curve Δ. Assume $n \geq n_o(\Delta)$: then the characteristic system of Δ is complete if and only if the characteristic system of Σ is complete.

This theorem, combined with Mumford's example 9.11 of a family of space curves Δ for which the characteristic system is never complete for each Δ, shows the existence of many surfaces Σ such that all their equisingular deformations do not have a complete characteristic system: in fact, given an n such that $\mathcal{O}_{\mathbb{P}^3}(n)(-2\Delta)$ is generated by global sections, it follows by Bertini's theorem that the general section $f \in H^o(\mathcal{O}_{\mathbb{P}^3}(n)(-2\Delta))$ defines a surface Σ smooth away from Δ, and with ordinary singularities only.

This result, obtained 20 years ago, culminated a very long history of attempts to show that the characteristic system of a surface Σ with ordinary singularities should always be complete (we defer the reader to [En], [Za], especially Mumford's appendix to chapter V for a more thorough discussion).

We simply want to remark again that the fact that the characteristic system is not com plete does not imply the singularity of the base B of the Kuranishi family: in fact, for $t \in B$ one can have a deformation φ_t of the holomorphic map $\varphi: S \to \mathbb{P}^3$ if and only if the cohomology class of H $(= \varphi^*$ (hyperplane)) remains of type $(1,1)$ on S_t.

Example 12.19. A classical case where Kodaira's theorem 12.17 applies is the case of Enriques' surfaces Σ, with equation

$$f(x_0,x_1,x_2,x_3) = x_0^2 x_1^2 x_2^2 x_3^2 \left(\sum_{i=0}^{3} \frac{1}{x_i^2} \right) + x_0 x_1 x_2 x_3\, q(x) ,$$

where $q(x)$ is a general quadratic form. Here Δ consists of the six edges of the coordinate tetrahedron $\{x_0 x_1 x_2 x_3 = 0\}$, $n = 6$. The normalization $\hat{\Delta}$ of Δ consists of 6 copies of $\mathbb{P}^1$, and $\mathcal{O}_{\hat{\Delta}}(H-T)$ has degree (-1) on each component, hence is non special $(H^1(\mathcal{O}_{\mathbb{P}^1}(-1)) = 0\,!)$. The surface Σ' to be taken is a general cubic surface of equation

$$x_0 x_1 x_2 x_3 \left(\sum_{i=0}^{3} a_i \frac{1}{x_i} \right) = 0 ,$$

hence $n_o(\Delta) \geq 6$, and Σ is regular.

The characteristic system has dimension 25 and it is easy to see that a smooth complete family of deformations of Σ is obtained by taking images under projectivities of surfaces in the above 10-dimensional family. Working out the exact sequence 12.6, we see that the above 10-dimensional family has bijective Kodaira-Spencer map, so that the Kuranishi family of S is smooth, 10-dimensional. This last result can also be gotten in a simpler way: since $K_S = 2H - D$, and $\varphi^*(x_0 x_1 x_2 x_3) = 2D$, we get $2K_S \equiv 0$. On the other hand, if $K_S \equiv 0$, there would be a quadric containing Δ, what is easily seen not to occur. Hence $K_S \not\equiv 0$, $2K_S \equiv 0$. Moreover $\chi(\mathcal{O}_S) = 1$. Taking the square root w of $x_0 x_1 x_2 x_3$ and then normalizing the surface $\Sigma' = \{(w, x_0, x_1, x_2, x_3) \mid w^2 = x_0 x_1 x_2 x_3,\ f(x) = 0\}$, we get a smooth surface S' (called a K3 surface) possessing an unramified double cover $\Pi: S' \to S$. It is easy to see that $K_{S'} \equiv 0$, and, since $\chi(\mathcal{O}_{S'}) = 2$, $H^1(\mathcal{O}_{S'}) = H^o(\Omega^1_{S'}) = 0$. Now $H^2(\Theta_S)$ is the Serre dual of $H^o(\Omega^1_S \otimes \Omega^2_{S'}) = H^o(\Omega^1_{S'}) = 0$, hence there are no obstructions for the Kuranishi family of S (of S', too).

Example 12.20 ([Ko 3], [Hor 4], [Us]). Let $\Delta \subset \mathbb{P}^3$ be a smooth curve, the complete intersection of two surfaces, $\Delta = \{F = G = 0\}$. Then one can consider the smooth family of surfaces of degree n having Δ as double curve. By our assumption, it is easy to check that, if $\deg F = a$, $\deg G = b$, the equation of Σ can be written in the form

(12.21) $$AF^2 + 2BFG + CG^2 ,$$

where $\deg A = n - 2a$, $\deg B = n - a - b$, $\deg C = n - 2b$.

The results of Kodaira-Horikawa and Usui can be summarized as follows: varying A, B, C one gets a surjective characteristic map, so that

(12.22 i) the characteristic system is complete.

Using the standard Euler sequence, it is possible to prove that, in the exact sequence

$$H^o(N'_{\Sigma|\mathbb{P}^3}) \to H^1(\Theta_S) \to H^1(\varphi^* \Theta_{\mathbb{P}^3}) \xrightarrow{\sigma} H^1(N'_{\Sigma|\mathbb{P}^3})$$

the homomorphism σ is injective, hence in particular

(12.22 ii) The Kuranishi family of S is smooth.

Furthermore,

(12.22 iii) the above surfaces are not (semi)-regular,

(12.22 iv) the pairing $H^1(\Theta) \times H^o(\Omega^2_s) \to H^1(\Omega^1_S)$ is non degenerate in the first factor (this result is called Infinitesimal Torelli property, and actually Usui proves the above result, provided $n \geq n_o(\Delta)$, also in the more general case of theorem 12.17).

This example shows clearly how the condition of semi-regularity is much too restrictive (in fact, as we noticed, it is an analogue of the condition $H^2(\Theta_X) = 0$ in order to ensure smoothness of the Kuranishi family).

§13. Generic multiple planes and equisingular deformations of plane curves with nodes and cusps

We consider again a smooth surface S and a finite morphism $\varphi: S \to \mathbb{P}^2$, of degree d; we let, as usual, H be the pull-back of a line, and we denote by R the ramification divisor of φ, i.e., given the exact sequence

(13.1) $$0 \to \Theta_S \xrightarrow{\varphi_*} \varphi^*\Theta_{\mathbb{P}^2} \to N_\varphi \to 0$$

the divisor of zeros of $(\Lambda^2\varphi_*) \in H^o((\Lambda^2\varphi^*\Theta_{\mathbb{P}^2}) \otimes (\Lambda^2\Theta_S)^\vee)$.

The following is a classical

Definition 13.2. φ is said to be a generic multiple plane (or a stable morphism) if R is smooth, $\varphi(R) = B$ (the branch locus) has only nodes and ordinary cusps as singularities, $\varphi|_R : R \to B$ is generically 1-1.

At the points of R, the normal form of φ (for suitable local holomorphic coordinates in the source and in the target) is as follows:

(13.3 i) $\varphi(x,y) = (x^2, y)$ at the points $p \in R$ with $\varphi(p)$ not a cusp of B

(13.3 ii) $\varphi(x,y) = (y, yx - x^3)$ at the points p_i of R with $\varphi(p_i) = p_i'$ a cuspidal point of B.

(13.3 i) and ii) imply that N_φ is an invertible $\mathcal{O}_R$-sheaf. It is easier though, to compute $\varphi_*(N_\varphi)$: we apply φ_* to 13.1 to obtain

$$0 \to \varphi_*\Theta_S \xrightarrow{\hat{\varphi}} \Theta_{\mathbb{P}^2} \otimes \varphi_*\mathcal{O}_S \to \varphi_* N_\varphi \to 0 .$$

Proposition 13.4. The trace map $t\colon \Theta_{\mathbb{P}^2} \otimes \varphi_*\mathcal{O}_S \to \Theta_{\mathbb{P}^2}$ induces an isomorphism of $\varphi_* N_\varphi$ with $\Theta_{\mathbb{P}^2}/\Theta_{\mathbb{P}^2}(-\log B)$; in particular, $\varphi_* N_\varphi = N'_{B|\mathbb{P}^2}$.

Proof. The statement is almost obvious for the points of $\mathbb{P}^2$ which are not cuspidal points of B. At a cuspidal point p_i', if $p_i \in R$ is such that $\varphi(p_i) = p_i'$, we see immediately that

$$(\varphi_* N_\varphi)_{p_i'} \cong \Theta_{\mathbb{P}^2} \otimes \varphi_*\mathcal{O}_{S,p_i} / \hat{\varphi}(\varphi_*\Theta_{S,p_i}) .$$

These two are $\mathcal{O}_{\mathbb{P}^2, p_i'}$ modules and, if we choose coordinates as in 13.3 ii), (y,z) at p_i such that $z = yx - x^3$, a basis for $\varphi_*\Theta_{S,p_i}$ is given by

$$\frac{\partial}{\partial x}, x\frac{\partial}{\partial x}, x^2\frac{\partial}{\partial x}, \frac{\partial}{\partial y}, x\frac{\partial}{\partial y}, x^2\frac{\partial}{\partial y},$$

while a basis for $\Theta_{\mathbb{P}^2} \otimes \varphi_*\mathcal{O}_{S,p_i}$ is given by

$$\frac{\partial}{\partial y}, x\frac{\partial}{\partial y}, x^2\frac{\partial}{\partial y}, \frac{\partial}{\partial z}, x\frac{\partial}{\partial z}, x^2\frac{\partial}{\partial z} .$$

Since the Jacobian matrix φ_* is $\begin{pmatrix} 0 & 1 \\ (y-3x^2) & x \end{pmatrix}$, we see that the image of $\hat{\varphi}$ is the submodule generated by

$$(13.5)\quad \begin{cases} \dfrac{\partial}{\partial y} + x\dfrac{\partial}{\partial z},\; x\dfrac{\partial}{\partial y} + x^2\dfrac{\partial}{\partial z},\; x^2\dfrac{\partial}{\partial y} + (yx-z)\dfrac{\partial}{\partial z}, \\[2ex] (y-3x^2)\dfrac{\partial}{\partial z},\; (3z-2xy)\dfrac{\partial}{\partial z},\; (3xz-2yx^2)\dfrac{\partial}{\partial z}. \end{cases}$$

We use now the symbol $\equiv$ to denote congruence modulo the submodule image $(\hat{\varphi})$, and we deduce from (13.5) that

$$x\frac{\partial}{\partial z} \equiv -\frac{\partial}{\partial y},\; x^2\frac{\partial}{\partial z} \equiv \frac{1}{3}y\frac{\partial}{\partial z},\; x\frac{\partial}{\partial y} \equiv x^2\frac{\partial}{\partial z} \equiv -\frac{1}{3}y\frac{\partial}{\partial z},$$

$$x^2\frac{\partial}{\partial y} \equiv z\frac{\partial}{\partial z} - yx\frac{\partial}{\partial z} \equiv z\frac{\partial}{\partial z} + y\frac{\partial}{\partial y},\; 3z\frac{\partial}{\partial z} \equiv 2yx\frac{\partial}{\partial z} \equiv -2y\frac{\partial}{\partial y},$$

$$0 \equiv 3zx\frac{\partial}{\partial z} - 2yx^2\frac{\partial}{\partial z} \equiv -3z\frac{\partial}{\partial y} - \frac{2}{3}y^2\frac{\partial}{\partial z}.$$

We readily infer that $\varphi_*(N_\varphi)$ is isomorphic to the quotient of $\Theta_{\mathbb{P}^2}$ by the submodule generated by $2y(\partial/\partial y) + 3z(\partial/\partial z)$ and by $9z(\partial/\partial y) + 2y^2(\partial/\partial z)$: it is immediate to check that this last submodule is indeed $\Theta_{\mathbb{P}^2}(-\log B)$, since the equation of $B = \varphi(R)$ is given by $4y^3 - 27z^2 = 0$. <u>Q. E. D.</u>

The previous proposition shows that infinitesimal deformations of the stable map φ correspond to infinitesimal equisingular deformations of B. On the other hand, if $\{\varphi_t : S_t \to \mathbb{P}^2\}_{t\in T}$ is a deformation of φ, it is easy to verify that the condition that φ_t be stable is an open one, and $\{B_t\} = \{\varphi_t(R_t)\}$ is an equisingular family of plane curves with nodes and cusps. Conversely, if $\{B_t\}_{t\in T}$ is an equisingular family of curves with nodes and cusps which is a deformation of $B = B_{t_o}$, we see that for t near to t_o the pairs $(\mathbb{P}^2, B)$ and $(\mathbb{P}^2, B_t)$ are diffeomorphic, and in particular $\Pi_1(\mathbb{P}^2 - B_t) \cong \Pi_1(\mathbb{P}^2 - B)$. Thus, the associated subgroup of the covering $\varphi: S - \varphi^{-1}(B) \to \mathbb{P}^2 - B$ determines another smooth surface S_t with a stable morphism $\varphi_t: S_t \to \mathbb{P}^2$ and it is not difficult to verify that in this way we get a deformation of φ with base T. We have thus

<u>Theorem 13.6</u>. There is a natural isomorphism between the characteristic system $H^o(N_\varphi)$ of a generic multiple plane φ and the equisingular characteristic system $H^o(N'_{B|\mathbb{P}^2})$ of its branch curve B. Also, the characteristic system of φ is complete if and only if the equisingular characteristic system of B is complete.

Again 9.11 and 12.18 imply the existence of plane curves with nodes and cusps whose equisingular characteristic system is obstructed: in fact, if $\varphi' : S \to \Sigma \subset \mathbb{P}^3$ is a map to a surface with ordinary singularities, it is well known that there exists a point p in $\mathbb{P}^3$ such that the projection $\Pi : \mathbb{P}^3 - \{p\} \to \mathbb{P}^2$ with centre p makes $\varphi = \Pi \circ \varphi'$ into a generic multiple plane. It is then clear that if $\{\varphi'_t\}$ is a deformation of φ', then $\{\varphi_t = \Pi \circ \varphi'_t\}$ is a deformation of φ. Conversely, as remarked before example 12.19, if $\{\varphi_t\}$ is a deformation of t, then there is a deformation φ'_t of φ' if and only if setting $H_t = \varphi^*_t$ (hyperplane in $\mathbb{P}^2$), the four sections $x_0, \ldots, x_3$ of $H^0(\mathcal{O}_S(H))$ extend holomorphically in t to 4 sections $x_{0t}, \ldots, x_{3t}$ of $H^0(\mathcal{O}_{S_t}(H_t))$.

This property holds in particular if $\dim H^1(\mathcal{O}_{S_t}(H_t))$ is independent of t: we defer the reader to [Wah] for more details, as well as for a very precise account of the theory of equisingular deformations of curves with nodes and cusps. Again here we have to remark that Enriques tried several times to show that curves with nodes and cusps were unobstructed, but this is not true, by the example of Mumford-Kodaira-Wahl.

LECTURE FIVE: SURFACES AND THEIR INVARIANTS

§14. Topological invariants of surfaces

In this section and the following ones, we shall very quickly review some basic facts about the topology of compact complex surfaces, and roughly outline the Enriques-Kodaira classification of (compact) complex surfaces. We defer the reader to [Bo-Hu], [Be 1], [B-P-V], and also to the survey papers [Ci], [Ca 2] for a thorough, update and exhaustive treatment. Given a (compact) complex surface S, we shall consider its underlying structure as an oriented topological 4-manifold, and also its differentiable manifold structure.

The main topological invariants of S are

$$(14.1)\quad \begin{cases} \Pi_1(S)\text{: the fundamental group of } S \\ b_i(S) = \dim_{\mathbb{Q}} H_i(S,\mathbb{Q}), \text{ the Betti numbers of } S \\ e(S) = \sum_{i=0}^{4} (-1)^i b_i(S) = 2 - 2b_1 + b_2, \text{ the topological Euler-Poincaré} \\ \qquad \text{characteristic of } S \\ T\text{: the torsion subgroup in } H_1(S,\mathbb{Z}) \text{ (and in } H^2(S,\mathbb{Z})) \\ Q = H^2(S,\mathbb{Z}) \to \mathbb{Z}, \text{ the integral unimodular quadratic form given by} \\ \qquad \text{cup product (composed with evaluation on the fundamental} \\ \qquad \text{class of } S) \\ b^+, b^-\text{: the indices of positivity, resp. negativity, of } Q \\ \tau = b^+ - b^-\text{: the signature of the manifold (note that the rank of } Q \\ \qquad \text{is } b_2 = b^+ + b^-) \end{cases}$$

The differentiable structure determines the real tangent bundle of S and its second Stiefel-Whitney class $w_2(S)$ (cf. [Mi-Sta]), by a theorem of Wu, determines whether $Q(x)$ is an even (i.e. $Q(x)$ even $\forall$ x in $H^2(S,\mathbb{Z})$) or odd form, since $Q(x) \equiv w_2(S)\cdot x \pmod 2$. Now, it is known (cf. [Se 2]) that all indefinite unimodular quadratic forms are determined by their rank, signature and party: if they are odd, then they are diagonalizable over $\mathbb{Z}$ (hence with ± 1 entries on the diagonal), and if they are even, they can be brought to a block diagonal form, with building blocks $U_2 = \begin{pmatrix} 0 & 1 \\ 1 & 0 \end{pmatrix}$ and

$$E_8 = \left(\begin{array}{cccc|cccc} 2 & 0 & -1 & 0 & & & & \\ 0 & 2 & 0 & -1 & & & & \\ -1 & 0 & 2 & -1 & & & & \\ 0 & -1 & -1 & 2 & -1 & & & \\ \hline & & & -1 & 2 & -1 & & \\ & & & & -1 & 2 & -1 & \\ & & & & & -1 & 2 & \end{array}\right) \quad \text{or } -E_8$$

Notice that $\tau(U_2) = 0$, $\tau(E_8) = 8$, and, by a theorem of Rokhlin [Rk], if $w_2 = 0$, then $\tau \equiv 0 \pmod{16}$ ($w_2 = 0 \Rightarrow Q$ is even, but not conversely, cf. [Hab] or 17.6).

What can be said about Q when Q is definite? Donaldson [Do 1] recently established the following remarkable result.

Theorem 14.2. Let M be a compact oriented 4-dimensional manifold with definite intersection form Q: then Q is diagonalizable (i.e., its matrix is $\pm$ Identity in a suitable basis).

The importance of the intersection form Q lies in the fact that it is the unique topological invariant when the 4-manifold M is simply connected. We have in fact ([Fre])

Theorem 14.3 (M. Freedman). Let M, M' be compact oriented topological 4-manifolds, and assume that they are simply-connected, and have the same intersection form Q. If M, M' have a differentiable structure, they are topologically equivalent. More generally, given Q, there are at most two topological types of 4-manifolds M with form Q and $\Pi_1(M) = 0$: if there are two, they are distinguished by the property whether $M \times [0,1]$ admits or doesn't admit a differentiable structure.

§15. Analytic invariants of surfaces

Before giving a list of invariants, it is convenient to clarify that in some cases we are talking about biholomorphic invariants, in others about bimeromorphic invariants. To explain the notion of a bimeromorphic map, we recall that two smooth algebraic varieties X and Y were classically said to be birational if their fields of rational functions $\mathbb{C}(X)$, $\mathbb{C}(Y)$ would be isomorphic. Such an isomorphism does not induce a biholomorphic map, but only a "generalized" graph, i.e., a closed subvariety Γ of $X \times Y$ such that, p_1, p_2 being the projections on both factors:

(15.1) i) there exist closed subvarities I_X of codimension at least 2 in X (resp.: I_Y), such that the restriction of p_1 from

$$\Gamma - p_1^{-1}(I_X) \to X - I_X$$

is biholomorphic (same condition for p_2),

ii) Γ is irreducible (hence Γ is the closure of $\Gamma - p_1^{-1}(I_X)$).

Replacing the word "subvariety" by the word "closed analytic subspace" (actually, in the sequel we shall make little distinction, since by Chow's theorem a closed analytic subspace of a compact algebraic variety is an algebraic subvariety), we obtain the definition of a bimeromorphic map.

Definition 15.2. A bimeromorphic map between compact complex manifolds X, Y is a biholomorphic map φ between open sets of X and Y, U_X and U_Y, such that $X - U_X$, $Y - U_Y$ are closed analytic subsets, and the closure Γ of the graph of φ is a closed analytic subset of $X \times Y$ satisfying properties 15.1.

In general, for a compact complex manifold X, we denote by $\mathbb{C}(X)$ its field of meromorphic functions and we recall the following famous result of Siegel [Sie 1].

Theorem 15.3. $\mathbb{C}(X)$ is a finitely generated extension field of $\mathbb{C}$ with transcendence degree $a(X)$ over $\mathbb{C}$ with $a(X) \leq n = \dim_{\mathbb{C}} X$.

Remark 15.4. It is easy to see that a bimeromorphic map between X and Y induces an isomorphism between $\mathbb{C}(X)$ and $\mathbb{C}(Y)$. Moreover, for an algebraic variety X $\mathbb{C}(X)$ coincides with the field of rational functions and, if $\dim Y = a(Y)$, $\mathbb{C}(Y) \cong \mathbb{C}(X)$, then Y is bimeromorphic to X. Such a Y does not need to be algebraic if $\dim \geq 3$, and is usually called a Moishezon manifold (cf. [Moi 1], [Moi 2]).

In dimension 2 the bimeromorphic maps are obtained as composition of certain elementary bimeromorphic maps which we are going now to describe.

Example 15.5. Let X be a complex manifold, p a point in X, $(z_1, \ldots, z_n)$ coordinates in a neighborhood U of p, with p corresponding to the origin. Let $\tilde{U}$ be the closure of the graph of the meromorphic map $U \to \mathbb{P}^{n-1}$ sending $(z_1, \ldots, z_n)$ to the line $\mathbb{C}(z_1, \ldots, z_n)$. Then, glueing $\tilde{U}$ with $X - \{p\}$ in an obvious way, we obtain a new manifold $\tilde{X}$, with a proper holomorphic and bimeromorphic map $\sigma: \tilde{X} \to X$ such that

i) $\sigma^{-1}(p)$, which is called the exceptional subvariety and denoted by E, is isomorphic to $\mathbb{P}^{n-1}$.

ii) The normal bundle $N_{E|\tilde{X}}$ is isomorphic to $\mathcal{O}_{\mathbb{P}^{n-1}}(-1)$.

iii) $\sigma|_{\tilde{X}-E}$ is a biholomorphism.

σ is called an elementary modification and σ^{-1} is called the blow-up of the point p.

The following result is classical.

Theorem 15.6. Every bimeromorphic map of complex surfaces factors as a composition of blow-ups followed by a composition of elementary modifications.

Definition 15.7. A complex surface S is said to be minimal if every holomorphic and bimeromorphic map $\sigma : S \to S'$ is a biholomorphism.

Remark 15.8. If $\tilde{S} \overset{\sigma}{\to} S$ is an elementary modification, then the lattice $H^2(\tilde{S}, \mathbb{Z})$, equipped with the quadratic form $\tilde{Q}$, is the orthogonal direct sum $H^2(S, \mathbb{Z}) \oplus \mathbb{Z}E$, where E is the cohomology class of the exceptional curve E, with $\tilde{Q}(E) = -1$. In particular $\tilde{Q}$ is odd and $b_2(\tilde{S}) = b_2(S) + 1$.

Conversely, if a complex surface $\tilde{S}$ contains an exceptional curve E of the I kind, i.e. $E \cong \mathbb{P}^1$, $N_{E|\tilde{S}} = \mathcal{O}_{\mathbb{P}^1}(-1)$ (or, equivalently $\tilde{Q}(E) = -1$) then there exists an elementary modification $\sigma : \tilde{S} \to S$ with $\sigma(E) =$ a point in P, and the second Betti number of S is equal to $b_2(\tilde{S}) - 1$. This is the classical result of Castelnuovo and Enriques, extended by Kodaira in the non algebraic case, and generalized by Grauert in [Gr 2]; combining this with another deep theorem of Castelnuovo and Kodaira, we obtain the following

Theorem 15.9. A complex surface S is minimal if and only if it does not contain an exceptional curve of the I kind. Every complex surface S' is a blow-up of a minimal surface S, and S is unique up to biholomorphism except if S' is ruled, i.e. S' is bimeromorphic to a product $C \times \mathbb{P}^1$.

Another biproduct of the structure theorem 15.6 of bimeromorphic maps of surfaces is the following theorem of Chow and Kodaira.

Theorem 15.10. A (smooth) complex surface S is projective (i.e., a submanifold of some projective space) if and only if $a(S) = 2$.

We should also remark that, by the result of Kodaira ([Ko 1], thm. 6) quoted at the end of §9, all small deformations of a non minimal complex surface are again non minimal, while we saw (use prop. 6.19, and the fact that $\mathbb{F}_1$ is the blow-up of a point in $\mathbb{P}^2$) that is is not true for deformations in the large.

We can now start to review some of the classical bimeromorphic invariants of surfaces. The following are numerical invariants.

(15.11)
$$\begin{cases} p_g \text{, the \underline{geometric genus}, is } h^2(\mathcal{O}_S) = \dim_{\mathbb{C}} H^2(\mathcal{O}_S) \\ q \text{, the \underline{irregularity}, is } h^1(\mathcal{O}_S) \\ P_m \text{, the \underline{m^{th} plurigenus}, is } h^o((\Omega^2_S)^{\otimes m}) = h^o(\mathcal{O}_S(mK)) \end{cases}$$

A more subtle invariant is the graded ring

(15.12)
$$\mathcal{R}(S) = \bigoplus_{m=0}^{\infty} H^o(\mathcal{O}_S(mK)), \text{ the \underline{canonical ring}.}$$

Definition 15.13. Let $Q(S)$ be the field of fractions of homogeneous elements of the same positive degree in $\mathcal{R}(S)$: then, either $Q(S) = \phi$, or $Q(S)$ is algebraically closed in $\mathbb{C}(S)$, and the Kodaira dimension of S, Kod(S) is

$$\begin{cases} -\infty & \text{if } Q(S) = \phi \\ \text{tr deg}_{\mathbb{C}} Q(S), & \text{otherwise.} \end{cases}$$

The above definition is not the unique possible one: denoting by φ_m the mth pluricanonical map, i.e. the rational map $\varphi_m : S \dashrightarrow \mathbb{P}^{P_m - 1}$ attached to the sections of $H^o(\mathcal{O}_S(mK))$, one can also define Kod(S) to be $\max_m (\dim \varphi_m(S))$.

From the complex point of view, one can consider divisors D, D' and consider the intersection number of two divisors, $D \cdot D'$, as classically defined through algebraic equivalence: since to a divisor D one associates the invertible sheaf $\mathcal{O}_S(D)$, one sees that, denoting by c_1 (I Chern class) the homomorphism of $H^1(\mathcal{O}^*_S)$ (classifying isomorphism classes of invertible sheaves) to $H^2(S, \mathbb{Z})$ appearing in the cohomology sequence of the exponential sequence

$$0 \to 2\pi i \mathbb{Z} \to \mathcal{O} \xrightarrow{\exp} \mathcal{O}^* \to 0 ,$$

the bilinear form of intersection on $H^2(S, \mathbb{Z})$ extends the classical intersection product. It is in fact true more generally that many of the analytical invariants, whose definition depends upon the complex structure of S, are in fact determined only by the topological structure of S.

Notice that, by 15.6 and 15.8, $\Pi_1(S)$, T, b_1, b^+ are bimeromorphic invariants.

Another classical invariant is

(15.14)
$$p^{(1)} = K^2 + 1, \text{ the linear genus of } S.$$

K^2, as well as $p^{(1)}$, are not bimeromorphic invariants, but, if a surface S' is not ruled, one can consider, e.g., the linear genus, and all the possible analytical and topological invariants of the unique minimal surface S bimeromorphic to S′ (S is called the minimal model of S′).

If you perform a blow-up, $e = 2 - 2b_1 + b_2$ goes up by 1, K^2 drops by 1: hence $K^2 + e$ is a bimeromorphic invariant, and an extension to complex surfaces of a classical theorem of Noether identifies it with a combination of previously encountered invariants. We have (cf. [Hi 1]), the following

Theorem 15.15. $(K^2 + e) = 12(1 - q + p_g) = 12\chi$ ($\chi = 1 - q + p_g$ is the Euler-Poincaré characteristic of the structure sheaf $\mathcal{O}_S$).

Another result of the same type, which represented a real breakthrough in the classification of complex manifolds, is the index theorem of Atiyah-Singer-Hirzebruch.

Theorem 15.16. $3\tau = 3(b^+ - b^-) = K^2 - 2e$.

Let's observe now that, by Serre duality, p_g is also the dimension of $H^o(\Omega^2_S)$; in general holomorphic 1 and 2 forms on a surface are d-closed, so that there is, using DeRham's theorem, an inclusion $H^o(\Omega^1_S) \subset H^1(S, \mathbb{C})$, $H^o(\Omega^2_S) \subset H^2(S, \mathbb{C})$. Moreover, if $\eta \in H^o(\Omega^1_S)$, $\bar{\eta}$ a $\bar{\partial}$-closed (0,1) form and gives, by Dolbeault's theorem, a cohomology class in $H^1(\mathcal{O})$. In this way one sees that $b^+ \geq 2p_g$, $h^o(\Omega^1_S) \leq q$, $(2q - b_1) \geq 0$, and the upshot is that, by a clever manipulation, the index theorem tells you that the sum (of positive integers!) $(b^+ - 2p_g) + (2q - b_i)$ equals 1: we thus have, (cf. [Ko 4]).

Theorem 15.17. If b_1 is even, $b_1 = 2q$, $b^+ = 2p_g + 1$, $h^o(\Omega^1_S) = q$; if b_1 is odd, $b_1 = 2q - 1$, $b^+ = 2p_g$, $h^o(\Omega^1_S) = q - 1$. In particular, if b_1 is even and $p_g = 0$, S is a projective surface.

There are several other results pertaining to inequalities between numerical invariants, or limitations in their range, but it is more convenient to postpone these to the next section, as being part of the classification theory.

We just end this paragraph by mentioning a consequence of the previous theorem

Corollary 15.18. The intersection form of a complex surface S is semi-negative definite if and only if b_1 is odd and $p_g = 0$.

LECTURE SIX: OUTLINE OF THE ENRIQUES-KODAIRA CLASSIFICATION

§16. Definition of the main classes of the classification

First of all, the main purpose of the Enriques-Kodaira classification is to partition all surfaces, considered only up to bimeromorphic equivalence, into 7 classes, in such a way that the knowledge (by explicit calculations) of some numerical invariant (or some more refined invariant, as the triviality of the canonical divisor) may allow to draw several conclusions about the structure and the geometry of some surfaces taken into consideration.

Its analogue in dimension 1 is the rough subdivision of curves according to the Kodaira-dimension, Kod = $-\infty$ if the genus g is 0, i.e. the curve is $\mathbb{P}^1$, Kod = 0 if $g = 1$, i.e. you have an elliptic curve, Kod = 1 if the genus is at least 2. The classification of curves according to their genus is more refined, but it is closely related to the knowledge of the topology of algebraic curves; in the surface case, a complete classification, less rough than the one given by Enriques and Kodaira, seems for the time being out of reach, due to the problem of classifying all the surfaces of general type.

Definition 16.1. A surface S is said to be of general type if $\mathrm{Kod}(S) = 2$, or, equivalently, if $Q(S) = \mathbb{C}(S)$ (cf. 15.13).

Definition 16.2. A surface S is said to be rational if it is bimeromorphic to $\mathbb{P}^2$, in particular a rational surface is ruled.

Definition 16.3. A surface S is said to be elliptic if it admits an elliptic fibration, i.e., a surjective morphism $f: S \to B$, with B a curve, and with the smooth fibres of f being elliptic curves.

Remark 16.4. If S is elliptic, $a(S) \geq a(B) = 1$; conversely, if $a(S) = 1$, there exists a curve B with $\mathbb{C}(S) \cong \mathbb{C}(B)$, and an elliptic fibration $f: S \to B$, such that all the curves of S are components of the fibres of f. If S is elliptic, then $\mathrm{Kod}(S) \leq 1$, and S is not of general type.

We can now pass to the list of the seven classes of surfaces, some of them being divided into subclasses.

Class 1): Ruled surfaces (i.e. bimeromorphic to a product $C \times \mathbb{P}^1$): these are distinguished by the irregularity q which equals the genus of C, and they are rational if $q = 0$. Their minimal models are $\mathbb{P}^2$, and $\mathbb{P}^1$-bundles over curves C.

From now on, since the minimal model is unique, we shall only talk about the minimal models.

Class 2): <u>K3 surfaces</u>, defined by the condition $q = 0$, $K \equiv 0$.

Class 3): <u>Complex tori</u>, i.e. surfaces biholomorphic to a quotient $\mathbb{C}^2/\Omega$, where Ω is a subgroup generated by 4 vectors linearly independent over $\mathbb{R}$.

Class 4): Elliptic surfaces with b_1 even, $P_{12} \neq 0$, $K \not\equiv 0$, divided into two subclasses distinguished by the Kodaira dimension.

Class 4), Kod = 0: <u>Enriques surfaces</u>, i.e. normalization of surfaces as in 12.19, or <u>hyperelliptic surfaces</u> (explicitly described in [B-DF], cf. [Be 1], pgs. 112-114), quotients of a product of two elliptic curves $E_1 \times E_2$ by the action of a subgroup G of E_1 acting on $E_1 \times E_2$ by sending (x_1, x_2) to $(x_1 + g, g(x_2))$, for a suitable action of G on E_2 such that $E_2/G \cong \mathbb{P}^1$.

Class 4), Kod = 1: Canonically elliptic surfaces with b_1 even: φ_{12}, the 12th pluricanonical map, gives an elliptic fibration.

Class 5) Surfaces of general type (cf. lecture seven).

Class 6) Elliptic surfaces with b_1 odd, $P_{12} \neq 0$, with subclasses

Class 6), Kod = 0: <u>Kodaira surfaces</u>, i.e. surfaces of the form $\mathbb{C}^2/G$, where G is a group of affine transformations of the form $(z_1, z_2) \mapsto (z_1 + a, z_2 + \bar{a} z_1 + b)$. They are distinguished into <u>primary</u> ones, with $b_1 = 3$, $K \equiv 0$, and <u>secondary</u> ones, with $b_1 = 1$, $K \not\equiv 0$; the secondary ones admit an unramified cover of finite degree which is a primary Kodaira surface.

Class 6), Kod = 1: Canonically elliptic surfaces with b_1 odd: φ_{12} gives an elliptic fibration.

Class 7): Surfaces with $b_1 = 1$, $P_{12} = 0$ (and Kod $= -\infty$ in fact): we shall not say much about these, since their classification has not been yet completely accomplished.

§17. Criteria of classification and features of some classes

The following result is the prototype of all criteria of classification

Theorem 17.1 (Castelnuovo's criterion). A surface S is rational if and only if $q = P_2 = 0$.

Ruled surfaces are also the ones which admit several characterizations.

Theorem 17.2. A surface S with b_1 even is ruled if and only if one of the following equivalent conditions is satisfied:

i) $P_{12} = 0$.

ii) These exists a curve C, not exceptional of the I kind, with $K \cdot C < 0$.

Moreover, if S is a minimal model and b_1 is even,

iii) $K^2 < 0$ if and only if S is ruled and $q \geq 2$.

The following results hold instead when S is minimal, but without the assumption that b_1 be even.

iv) $K^2 > 0$, $P_2 = 0$ if and only if S is rational

v) $K^2 > 0$, $P_2 \neq 0$ if and only if S is of general type.

vi) $e < 0$ if and only if S is ruled and $q \geq 2$.

The definition of K3 surfaces we gave in §16 was one of the less explicit ones: in fact, we could have chosen to define a K3 surface according to the following beautiful theorem of Kodaira (conjectured earlier by Andreotti and Weil).

Theorem 17.3. A minimal surface S is a K3 surface if and only if S is a direct deformation, with non singular base, of a non singular surface of degree 4 in $\mathbb{P}^3$.

We notice that some of the K3 surfaces can be elliptic, as well as some complex tori, and this fact justifies the condition $K \not\equiv 0$ used to define Class 4). We defer the reader to [B-P-V] for an excellent survey about K-3 surfaces and their moduli space.

The following theorem characterizes complex tori.

Theorem 17.4. A minimal surface S is a complex torus if and only if $b_1 = 4$, $K \equiv 0$. Moreover, a surface with $K \equiv 0$ (necessarily minimal) has $q = 0, 2$, and thus $b_1 = 0, 4, 3$ according to whether S is a K3 surface, a complex torus, or a primary Kodaira surface.

As to class 4), we recall that we have already implicitly remarked that the two subclasses are distinguished by the value of P_{12} being 1 or ≥ 2, we have in fact the following.

Proposition 17.5.

$\mathrm{Kod}(S) = -\infty$ if and only if $P_{12} = 0$.

$\mathrm{Kod}(S) = 0$ if and only if $P_{12} = 1$.

$\mathrm{Kod}(S) = 1$ if and only if $P_{12} \geq 2$ and $K^2 = 0$ for the minimal model of S.

In particular, for the subclass of class 4), where Kod = 0, we have a rather nice characterization.

Theorem 17.6. A surface S is hyperelliptic if and only if $P_{12} = 1$, $b_1 = 2$. A surface with $P_{12} = 1$, $b_1 = 0$ is a K3 surface if $p_g = 1$ and an Enriques surface if $p_g = 0$ (then $P_2 = 1$).

For the subclass of class 4) with Kod = 1, we see from proposition 17.5 that a characterization of a minimal model S is

(17.7) $$K^2 = 0,\ P_{12} \geq 2,\ b_1 \text{ even}$$

but, if we are given a non minimal surface, then the conditions b_1 even, and $|12K|$ yielding a rational map with image a curve are easier to check. As a matter of fact, to detect the exceptional curves of the I kind on a surface with $\mathrm{Kod} \geq 0$ (i.e. $P_{12} \neq 0$), the standard way is to look at the smallest m such that $P_m \neq 0$, and then to look at the fixed part of $|mK|$, to check which components of this divisor are exceptional.

Surfaces of general type being taken into account by theorem 17.2.v), we notice that the surfaces in classes 6) and 7) are the ones with b_1 odd and, in particular, they cannot be Kählerian. From surface classification and results of Kodaira, Miyaoka, Todorov and Siu follows also the remarkable

Theorem 17.8. A complex surface with b_1 even is a deformation of an algebraic surface and is Kählerian.

The two classes 6) and 7) are distinguished by the value of P_{12}, which is $\neq 0$ for class 6), and 0 for class 7) (remark, though, that Kodaira's class VII is different from our class 7), being defined by the condition $b_1 = 1$, and thereby including also secondary Kodaira surfaces and some canonically elliptic surfaces). The subclass of Kodaira surfaces is again characterized by

(17.9) $$P_{12} = 1, \ b_1 \text{ odd}$$

and we defer the reader to [Ko 4] for a very detailed description of these surfaces.

For lack of time, we don't attempt to describe the known examples and the classification results for surfaces in class 7), referring to [Nak 1] for a nice and updated survey.

Let's just notice that, when $b_1 = 1$, then, by a result of Kodaira on elliptic surfaces, we have $p_g = 0$, and the intersection form Q is semi-negative definite by 15.18. More precisely, by Noether's formula 15.15, since $\chi = 0$, $b_2^- = b_2 = e = -K^2$. Hence, as in the case of ruled surfaces with $q \geq 2$, $K^2 < 0$ as soon as the Betti number b_2 is $\neq 0$. On the other hand, if S is elliptic, by Kodaira's canonical divisor formula (cf. [B-P-V], pgs. 161-164), a multiple mK of K is linearly equivalent to a multiple rF of a fibre F of the elliptic fibration, hence in particular $K^2 = 0$ (the above canonical divisor formula shows also that a surface S admits more than one elliptic fibration only if it is not canonically elliptic and in fact only if Kod = 0, since either r = 0 or K determines the elliptic fibration).

We have thus the following

<u>Theorem 17.10</u>. A minimal surface S has $K^2 < 0$ if and only if either S is ruled with $q \geq 2$ (b_1 odd) or S has $b_1 = 1$, $b_2 > 0$.

Since $\chi = K^2 + e$, by Castelnuovo's theorem 17.2.vi) and the above one follows

<u>Theorem 17.11</u>. A surface S has $\chi < 0$ if and only if S is ruled with $q \geq 2$.

LECTURE SEVEN: SURFACES OF GENERAL TYPE AND THEIR MODULI

§18. Surfaces of general type, their invariants and their geometry

Let's observe that Noether's theorem 15.15 and the index theorem 15.16 ensure that the analytically defined invariants K^2 and χ are determined by e and τ, therefore K^2 and χ are topological invariants and the advantage of dealing with them stems from the fact that, unlike K^2 and e, they don't have to satisfy any congruence relation. Also, by theorem 17.2, we have $K^2, \chi \geq 1$ for a minimal surface S of general type (these are called Castelnuovo's inequalities), and two more inequalities are satisfied.

$$\text{(18.1)} \qquad K^2 \geq 2p_g - 4 \geq 2\chi - 6 \qquad \text{(Noether's inequality)}$$

$$K^2 \leq 9\chi \qquad \text{(Bogomolov-Miyaoka-Yau's inequality)}$$

In fact, by classification, the inequality is true for all surfaces, except for ruled surfaces of irregularity $q \geq 2$, which have $K^2 = 8(1-q)$, $\chi = 1 - q$. S. T. Yau [Ya] proved indeed a much stronger theorem, in particular it follows from his results the following

Theorem 18.2. If a surface of general type S has $K^2 = 9\chi$, then the universal cover $\tilde{S}$ of S is biholomorphic to the unit ball in $\mathbb{C}^2$.

An easily proven but nice corollary concerns the surfaces S for which the intersection form Q is positive definite: in fact, if $b_2 = b^+ = \tau$, since $K^2 \leq 9\chi$, $4K^2 = 3(12\chi) = 3(K^2+e)$, i.e., $K^2 - 2e \leq e$, which is in turn equivalent to $3\tau \leq e = 2 + b_2 - 2b_1$. Then $2b_2 \leq 2(1 - b_1)$ and thus $b_2 = 1$, $b_1 = 0$, $p_g = 0$ ($b^+ = 1 = 1 + 2p_g$) hence $\chi = 1$, $K^2 = 12 - 3 = 9$.

Corollary 18.3. The only complex surfaces for which the intersection form Q is positive definite have $b_2 = 1$ and $K^2 = 9$, $\chi = 1$: they are either $\mathbb{P}^2$ or a surface of general type with the unit ball as universal cover.

By a result of Kodaira, the plurigenera are completely determined by the invariants K^2, χ; we have in fact

Theorem 18.4. If S is a minimal surface of general type, and $m \geq 2$, $P_m = \chi + (1/2) K^2 m (m - 1)$.

As a consequence of the theory of pluricanonical mappings, that we are going now to explain, there follows the result that surfaces with given invariants K^2, χ belong to a finite number of deformation types.

Let's go back to the canonical ring $\mathcal{R}(S)$, defined in 15.12; 18.4 tells us that its Hilbert polynomial is determined by K^2, χ, and clearly two minimal surfaces of general type S, S' are isomorphic if and only if $\mathcal{R}(S)$ and $\mathcal{R}(S')$ are isomorphic graded rings. Before mentioning directly how to recover S from $\mathcal{R}(S)$, let's remark that, $\mathbb{C}(S)$ being finitely generated, there exists an m such that every function in $\mathbb{C}(S)$ can be written as a fraction whose numerator and denominator are sections of $H^0(\mathcal{O}_S(mK))$. In other terms, there exists an m such that the mth pluricanonical map φ_m is birational onto its image Σ_m. Unfortunately, unlike the case of curves, one cannot expect φ_m to be an embedding: in fact, though $K \cdot C \geq 0$ for each irreducible curve on S (17.2.ii)), there can be a finite number of curves $C \cong P^1$ with $K \cdot C = 0$ ($C^2 = -2$), and K is ample iff these curves do not exist on S. Otherwise, since $\mathcal{R}(S)$ is finitely generated, one can take $X =$ Proj $\mathcal{R}(S)$ (its points correspond to maximal homogeneous ideals in $\mathcal{R}(S)$), and there exists a holomorphic map $\Pi: S \to X$ satisfying the following properties

(18.5) i) X is a normal surface

ii) if $i: X^o \to X$ is the inclusion morphism of the non-singular part of X, then the sheaf of Zariski differentials $\omega_X = i_*(\Omega^2_{X^o})$ is invertible and $\Pi^*(\omega_X) = \mathcal{O}_S(K_S)$ (i.e., X has only Rational Double Points, R.D.P.'s, as singularities).

iii) every pluricanonical map $\varphi_m: S \to \Sigma_m$ factors through Π and $\tilde{\varphi}_m: X \to \Sigma_m$.

Definition 18.6. $X = \text{Proj}(\mathcal{R}(S))$ is called the canonical model of S.

We defer the reader to [Ca 3] for a survey of recent results on pluricanonical maps of surfaces of general type, and we content ourself with stating a by now classical result of Bombieri.

Theorem 18.7. If $m \geq 5$, $\tilde{\varphi}_m: X \to \Sigma_m$ is an isomorphism.

§19. Pluricanonical images and Gieseker's moduli variety

Recall that a projective variety $\Sigma \subset P^N$ is said to be projectively normal if the restriction homomorphisms

$$H^o(\mathcal{O}_{P^N}(k)) \to H^o(\mathcal{O}_\Sigma(k))$$

are surjective for each integer $k \geq 0$. In the case where $\Sigma = \Sigma_m$ is the pluricanonical image of a surface S of general type we have (cf. [Ci 2]).

Theorem 19.1. If $m \geq 8$, $\tilde{\varphi}_m : X \to \Sigma_m$ is an isomorphism onto a projectively normal surface.

We are going now to discuss very loosely the main line of ideas which lead to Gieseker's theorem about the existence of moduli spaces of surfaces of general type. First of all, let's recall Mumford's definition (cf. [Mu 2]).

Definition 19.2. A variety $\mathcal{M}_{K^2,\chi}$ is said to be a coarse moduli space for surfaces of general type S with given invariants K^2, χ if there exists a bijection λ between $\mathcal{M}_{K^2,\chi}$ and the set of isomorphism classes [S] of minimal surfaces as above, satisfying the following property: for each deformation $\mathcal{S} \xrightarrow{p} B$ of such surfaces, there is given a unique morphism $\psi : B \to \mathcal{M}_{K^2,\chi}$ such that $\psi(b) = \lambda^{-1}([S_b])$, and the correspondence $(\mathcal{S} \xrightarrow{p} B) \to \psi$ is compatible with pull-backs (i.e., to the family $f^*\mathcal{S} \to B'$ corresponds $\psi \circ f = \psi' : B' \to \mathcal{M}_{K^2,\chi}$).

Theorem 19.3 (Gieseker). There exists a coarse moduli space $\mathcal{M}_{K^2,\chi}$ which is a quasi-projective variety. Moreover, two surfaces S, S′ correspond to points in the same connected component of $\mathcal{M}_{K^2,\chi}$ if and only if S is a deformation of S′.

The key point consists in taking all the mth canonical images Σ_m of our surfaces (with K^2, χ fixed): they are, if $m \geq 8$, projectively normal surfaces of fixed degree ($= m^2K^2$) in a fixed projective space $\mathbb{P}$ of dimension $P_m - 1 = \chi - 1 + m(m-1)/2\, K^2$.

Now, Σ_m and Σ'_m are projectively equivalent if and only if the corresponding surfaces S, S′ are isomorphic, and one has to construct a quotient by the group $PGL(P_m)$ of a variety $\hat{\mathcal{H}}$ parametrizing our surfaces Σ_m. To this purpose, one has first to use the Hilbert scheme (cf. [Gro 2], [Mu 3], [Ser 2]) technique: since these notes are meant to be elementary, let's indicate the main idea.

Since $\Sigma = \Sigma_m$ is projectively normal, if we denote by $\mathcal{I}_\Sigma$ the ideal sheaf of Σ in $\mathbb{P}$, the space $H^o(\mathcal{I}_\Sigma(n))$ has fixed dimension equal to

$$\binom{n+P_m}{P_m} - P_{nm}$$

Now, there exists an integer n, depending only on the Hilbert polynomial P(n) of a variety (here $P(n) = \chi + K^2/2\ nm(nm-1)$, hence P depends only upon K^2, χ, m), such that the ideal sheaf $\mathcal{I}_\Sigma$ of Σ is generated by the subspace $V_\Sigma = H^o(\mathcal{I}_\Sigma(n))$ of the fixed vector space $W = H^o(\mathcal{O}_{\mathbb{P}}(n))$. Let r be the dimension of V_Σ, t the dimension of W: then $r = t - P(n)$, and r, t depend only upon K^2, χ, m.

Definition 19.4. The nth Hilbert point of Σ is the point $\mathbb{P}(\Lambda^r V_\Sigma) \in \mathbb{P}(\Lambda^r W)$, belonging to a Grassman manifold $G(r, W)$ of dimension $r \cdot P(n)$.

The Hilbert scheme of subschemes of $\mathbb{P}$ with Hilbert polynomial $\mathbb{P}(n)$ is the closed subscheme $\mathcal{H}$ of the Grassmann manifold $G(r, W) \subset \mathbb{P}(\Lambda^r W)$ corresponding to the set of r-dimensional subspaces V of W such that the ideal sheaf $\mathcal{I} = V\mathcal{O}_{\mathbb{P}}$ generated by V defines a subscheme $\Sigma(V)$ with Hilbert polynomial P(n). $\mathcal{H}$ is the basis of a universal family, i.e., there is a subscheme Z of $\mathcal{H} \times \mathbb{P}$ such that the fibre of Z over $V \in \mathcal{H}$ is just the subscheme $\Sigma(V)$. This $\mathcal{H}$ is too big, and first of all one has to take the open set $\mathcal{H}_o \subset \mathcal{H}$

(19.5) $\mathcal{H}_o = \{ V \in W \mid V\mathcal{O}_{\mathbb{P}}$ defines a connected surface $\Sigma(V)$ with only R. D. P.'s as singularities $\}$.

Over $\mathcal{H}_o$ lies the restriction Z_o of the universal family Z, and inside $\mathcal{H}_o$ lies the closed subscheme

(19.6) $\hat{\mathcal{H}} = \{ V \mid \Sigma = \Sigma(V)$ has $\omega_\Sigma(-1) \cong \mathcal{O}_\Sigma \}$.

The restriction $\hat{Z}$ of the universal family enjoys the following

19.7. Universal property of $\hat{Z} \subset \hat{\mathcal{H}} \times \mathbb{P}$: for each family $p: \mathcal{S} \to B$ of minimal surfaces of general type, with given invariants K^2, χ, and for each choice of P_m independent sections of the locally free sheaf $p_*(\omega_{\mathcal{S}|B}^{\otimes m})$, there does exist a unique pair of morphisms

$$\alpha: \mathcal{S} \to \hat{Z}, \quad f: B \to \hat{\mathcal{H}}$$

such that the following diagram commutes

$$\begin{array}{ccc} \mathcal{S} & \xrightarrow{\ \alpha\ } & \hat{Z} \\ {\scriptstyle p}\downarrow & & \downarrow \\ B & \xrightarrow{\ f\ } & \hat{\mathcal{H}} \end{array}$$

and such that α gives fibrewise the mth-canonical map

$$\varphi_m^{\cdot} : S_b \to \mathbb{P} \ .$$

Now, all the surfaces of general type with those given invariants K^2, χ appear as (minimal resolutions of) fibres of the family $\hat{Z} \to \hat{\mathcal{H}}$ of canonical models; since the base $\hat{\mathcal{H}}$ is quasi-projective, hence it has a finite number of components, and, by a result of G. Tjurina [Tju], the canonical models X, X′ of two surfaces of general type S, S′ are a deformation of each other if and only if S, S′ are deformations of each other, it is thus proven (cf. 19.3).

(19.8) The surfaces of general type with given K^2, χ belong to a finite number of deformation classes. In particular there is only a finite number of diffeomorphism types.

It is also clear that the group

(19.9) $G' = PSL(P_m)$ acts on $\hat{\mathcal{H}}$, and there is a bijection between the set of orbits of G and the set of isomorphism classes of surfaces of general type with invariants K^2, χ.

Thus, the problem of the existence of the coarse moduli space $\mathcal{M}_{K^2,\chi}$ is reduced to the existence of a categorical quotient (cf. [Mu 2], chap. I) for the action of G on $\hat{\mathcal{H}}$. This is a problem in the realm of the so-called Geometric Invariant Theory: in general (cf. [Mu 2], Appendix) such quotients always exist as algebraic (or Moishezon) spaces (i.e., as complex spaces bimeromorphic to algebraic varieties). In fact, we recall again, what Gieseker proves is that

(19.10) The categorical quotient $\hat{\mathcal{H}}/G'$ exists as a quasi-projective variety.

The idea is as follows, the Hilbert point belongs to

$$\mathbb{P}(\Lambda^r W) = \mathbb{P}(\Lambda^r (\mathrm{Sym}^n(\mathbb{C}^{P_m}))) \quad ,$$

and one wants to show that the invariant homogeneous functions of degree ℓ, i.e. the sections in

$$A = H^{o}(P(\Lambda^{r}(\mathrm{Sym}^{n}(\mathbb{C}^{P_m}))), \mathcal{O}(\ell))^{G'},$$

for m, n, ℓ sufficiently big separate the orbits, so that $\hat{\mathcal{K}}/G'$ sits inside a projecttive variety. We have here the typical situation of Geometric Invariant Theory: a vector space

$$U = \Lambda^{r}(\mathrm{Sym}^{n}(\mathbb{C}^{P_m})),$$

and an action of $SL(P_m) = G$. Then

(19.11) A point $u \in U$ is

1) unstable if $\overline{Gu} \ni 0$
2) semistable if $\overline{Gu} \not\ni 0$
3) stable if it is semistable and the stabilizer of u is finite.

and then the G invariant polynomials define a morphism ψ from $\mathbb{P}(U)^{ss}$ = {semi-stable points} to a projective variety Y in such a way that the restriction of ψ to the stable points $\mathbb{P}(U)^{s}$ separates the orbits (in fact, the stable points have closed orbits, and two closed orbits are separated by some G invariant polynomials.

Remark 19.12. In our case the condition that the stabilizer of u be finite, when u corresponds to the Hilbert point of a pluricanonical image Σ_m follows from a general result of Matsumura [Mat]. In fact, $\{g \mid g(\Sigma_m) = \Sigma_m\} = \mathrm{Aut}(\Sigma)$ is a linear algebraic group which, if not finite, would have a non-trivial Cartan subgroup, which is a rational variety: but then Σ would be uniruled (X, with dim X = n is said to be uniruled if there exist Y, with dim Y = n-1, and a dominant rational map of $Y \times \mathbb{P}^1$ into X), and in particular all the plurigenera of Σ would vanish.

The really difficult point is to prove that these Hilbert points are semistable, and this is done with hard combinatorial estimates using the Hilbert-Mumford stability criterion.

19.13. u is semi-stable if and only if for each 1-parameter subgroup of G, $t \in \mathbb{C}^* \to g(t)$, where

$$g(t) = A \begin{pmatrix} t^{a_o} & & 0 \\ & \ddots & \\ 0 & & t^{a_p} \end{pmatrix} A^{-1} \quad (\text{with } \Sigma a_i = 0)$$

one has $\lim_{t \to 0} g(t)(u) \neq 0$.

From the fact that $\hat{\mathcal{K}}/G = \mathcal{M}_{K^2,\chi}$ is a coarse moduli space (cf. 19.2) it follows easily the following

Corollary 19.14. Let S be a minimal surface of general type with invariants K^2, χ: then, if B is the base of the Kuranishi family of S, then, locally around the point [S] corresponding to the isomorphism class of S, $\mathcal{M}_{K^2,\chi}$ is analytically isomorphic to the quotient of B by the finite group Aut(S).

§ 20. The number of moduli M(S) of a surface

Definition 20.1. For a surface of general type S, we define M(S), the number of moduli of S, to be equal to the dimension of the base B of its Kuranishi family, (i.e. the maximum of the dimensions of the irreducible components of B.

Remark 20.2. By 19.14, M(S) is the dimension of $\mathcal{M}_{K^2,\chi}$ at the point [S] corresponding to S.

Remark 20.3. More generally, Kodaira and Spencer ([K-S], Chap. V, §11) define the number of moduli of a complex manifold X to be the maximal dimension of the base T of an effectively parametrized family $\mathcal{X} \to T$ of deformations of X, i.e. such that the Kodaira-Spencer map ρ_t is injective for each t.

It was conjectured by Noether that M(S) would be $10\chi - 2K^2$, whereas Enriques realized that $10\chi - 2K^2$ was only a lower bound for M(S). In fact, by the Hirzebruch-Riemann-Roch formula (cf. [Hi]), $-(10\chi - 2K^2)$ equals the Euler-Poincaré characteristic of Θ, i.e. $h^0(\Theta) - h^1(\Theta) + h^2(\Theta)$. In general, $H^0(\Theta_X)$ is the Lie algebra of a real Lie group of biholomorphisms of X: since (cf. 19.12) Aut(S) is finite, $H^0(\Theta_S) = 0$, and, as we saw in 3.3.

(20.4) $$10\chi - 2K^2 = h^1(\Theta) - h^2(\Theta) \leq \dim B = M(S).$$

On the other hand, $M(S) \leq h^1(\Theta_S)$, and we indeed conjecture that for the general surface S in each component of the moduli space $\mathcal{M}_{K^2,\chi}$ equality holds, provided $q(S) = 0$ and K is ample (for reasons stemming from [Ca 4] and [Ca 5]), so that $\mathcal{M}$ should be a reduced variety. In any case, finding the dimension $h^1(\Theta)$ is by no means easier (and in some cases even more difficult) than to compute M(S), therefore we just observe that $h^2(\Theta_S) = h^0(\Omega^1_S \otimes \Omega^2_S)$, hence

(20.5) $$M(S) \le h^1(\Theta) = 10\chi - 2K^2 + h^o(\Omega^1 \otimes \Omega^2) ,$$

and one can give an upper bound for $M(S)$ by giving an upper bound for $h^o(\Omega^1 \otimes \Omega^2)$ in terms of χ, K^2, q. It is clear that, doing so, one does not obtain the best estimate, because one is giving an upper bound for the dimension of the Zariski tangent space of each point of the base B of the Kuranishi family, and not simply an upper found for dim B, as we already noticed.

A way to bound $h^o(\Omega^1_S \otimes \Omega^2_S) = h^o(\Omega^1_S(K))$ is to use the existence of a smooth curve C in $|2K|$ as soon as $K^2 \ge 5$, or $p_g \ge 1$, except possibly if $K^2 = 3, 4$ (this follows from recent results of Francia [Fra] and Reider [Rei 2]). In fact then, by the exact sequences

$$0 \to \Omega^1_S(-K) \to \Omega^1_S(K) \to \Omega^1_S(K) \otimes \mathcal{O}_C \to 0$$

$$0 \to \mathcal{O}_C(-K) \to \Omega^1_S(K) \otimes \mathcal{O}_C \to \mathcal{O}_C(4K) \to 0$$

since $\Omega^1_S(-K) \cong \Theta_S$, we get

$$h^o(\Omega^1_S(K)) \le h^o(\mathcal{O}_C(4K)) = 5K^2 .$$

Otherwise, one looks for a smooth curve in $|mK|$, with $m = 3$, and the result is (with an improvement only of the constant in [Ca 6], Thm. B).

Theorem 20.6. We have the inequalities

$$10\chi - 2K^2 \le M(S) \le 10\chi + 3K^2 + 18 .$$

If $|K|$ contains a smooth curve, then $M(S) \le 10\chi + q + 1$.

In the case when the surface S has $q \ne 0$, since $q = h^o(\Omega^1_S)$, surely $h^o(\Omega^1_S \otimes \Omega^2_S) \ne 0$, since there is a bilinear map $H^o(\Omega^1_S) \times H^o(\Omega^2_S) \to H^o(\Omega^1_S \otimes \Omega^2_S)$ which is non-degenerate in each factor, nevertheless the existence of holomorphic 1-forms can be used for our desired bound.

For irregular ($q \ne 0$) surfaces, a powerful tool is given by the analysis of the Albanese map

$$\alpha : S \to A = \mathrm{Alb}(S) = H^o(\Omega^1_S)^\vee / H_1(S, \mathbb{Z}) ,$$

such that $\alpha(p) = \int_{p_o}^{p}$ (where p_o is a fixed point and the linear functional on $H^o(\Omega^1)$ is clearly defined only up to $\int_\gamma$, for $\gamma \in H_1(S, \mathbb{Z})$). The condition that

α should have differential of maximal rank = 2 except that at a finite set of points can also be phrased as

(20.7) there do exist $\eta_1, \eta_2 \in H^0(\Omega^1_S)$ such that $C = \mathrm{div}(\eta_1 \wedge \eta_2)$ is a reduced and irreducible curve (in $|K|$!).

Castelnuovo [Cas] tried to prove the inequality $M \le p_g + 2q$ under the assumption that $\alpha(S)$ would be a surface, i.e., assuming that the differential of α would have rank 2 outside of a curve: we showed in [Ca 6] that there are infinitely many families of surfaces S for which $\alpha(S)$ is a surface, and $M \ge 4p_g + o(p_g)$. In fact, we also proved that (20.7) implies $M \le p_g + 3q - 3$, and conjectured that (20.7) would imply the Castelnuovo inequality. We shall now give Reider's simple proof [Rei 1] of this inequality.

Theorem 20.8 (Reider). If S satisfies 20.7, then $M \le p_g + 2q$.

Proof. Since C is irreducible, η_1 and η_2 don't vanish simultaneously on a curve, hence we can assume $\eta = \eta_1$ to have only isolated zeros. Let Z be the 0-dimensional scheme of zeros of η: we have then the Koszul complex

$$0 \to \mathcal{O}_S \xrightarrow{\eta} \Omega^1_S \xrightarrow{\wedge\eta} \mathcal{I}_Z \Omega^2_S \to 0 \tag{20.9}$$

where $\mathcal{I}_Z$ is the ideal sheaf of Z. Tensoring with $\mathcal{O}_S(K)$, since $h^1(\Theta) = h^1(\Omega^1_S(K))$, we get

$$h^1(\Theta_S) \le q + h^1(\mathcal{I}_Z \Omega^2_S) - 1 . \tag{20.10}$$

The basic fact is that the ideal sheaf of C is contained in $\mathcal{I}_Z$, hence we have an exact sequence

$$0 \to \mathcal{O}_S(-C) \to \mathcal{I}_Z \to \mathcal{F} \to 0 , \tag{20.11}$$

where $\mathcal{F}$ is a torsion free, rank 1 sheaf on C. Tensoring (20.11) with $\mathcal{O}_S(2K)$ yields

$$h^1(\mathcal{I}_Z(2\ _S)) \le q + h^1(\mathcal{F}(2\ _S)) - 1 . \tag{20.12}$$

Since $\mathcal{O}_C(2K_S)$ is the dualizing sheaf of C, by Grothendieck duality

$$h^1(\mathcal{F}(2K_S)) = \dim \mathrm{Hom}(\mathcal{F}(K_S), \mathcal{O}_C(K)) .$$

On the other hand, there is a bilinear map

$$H^o(\mathcal{F}(K)) \times \mathrm{Hom}(\mathcal{F}(K),\ \mathcal{O}_C(K)) \to H^o(\mathcal{O}_C(K))$$

which is non-degenerate in each factor; since these are complex vector spaces, a well known application of the Segre product (projectived tensor product), gives the inequality

$$h^1(\mathcal{F}(2K_S)) \le h^o(\mathcal{O}_C(K)) - h^o(\mathcal{F}(K)) + 1$$

$$\le p_g + q - 1 - h^o(\mathcal{F}(K)) + 1 .$$

To give a lower bound for $h^o(\mathcal{F}(K))$, we tensor (20.11) with $\mathcal{O}_S(K)$, to get $h^o(\mathcal{F}(K)) \ge h^o(\mathcal{I}_Z(K)) - 1 \ge q - 2$ by the sequence (20.9). It suffices to put these inequalities together.

Q.E.D.

We defer the reader to [Rei 1] for other results of this type, and we note in fact that Reider simply uses the existence of a 1-form η with isolated zeros, and the existence of a reduced irreducible curve C in $|K|$ such that $\mathcal{O}_S(-C) \subset \mathcal{I}_Z$, so that the method can be generalized taking $C \in |mK|$ with such a property. It would be interesting to give an upper bound for $M(S)$ in the case of a surface S fibred over a curve B of genus ≥ 1.

LECTURE EIGHT: BIHYPERELLIPTIC SURFACES AND PROPERTIES OF THE MODULI SPACES

§21. Moduli spaces of surfaces of general type and their properties

Let S be a minimal surface of general type with invariants K^2, χ and consider the Gieseker moduli space $\mathfrak{M}_{K^2,\chi}$, which has a finite number of connected components. We can define two moduli spaces $\mathfrak{M}^{top}$ and $\mathfrak{M}^{diff}$ which are contained in $\mathfrak{M}_{K^2,\chi}$, and are indeed a union of connected components of $\mathfrak{M}_{K^2,\chi}$, as follows ($\mathfrak{M}^{diff} \subset \mathfrak{M}^{top}$).

Definition 21.1. $\mathfrak{M}^{top}(S)$ (resp.: $\mathfrak{M}^{diff}(S)$) is $\{[S'] \in \mathfrak{M}_{K^2,\chi} \mid$ there exists an orientation preserving homeomorphism (resp.: diffeomorphism) between S and S'$\}$.

We note (cf. §14.17) that if Q is even and $K^2 \geq 9$, then every complex structure on the topological manifold underlying S corresponds to a minimal surface S' of general type homeomorphic to S. Thus $\mathfrak{M}^{diff}$, e.g., is then a coarse moduli space for all the (integrable almost) complex structures on the differentiable manifold underlying S.

Remark 21.2. Since the Stiefel-Whitney class w_2 (cf. 14) is the mod 2 reduction of $c_1(K) \in H^2(S, \mathbb{Z})$, we see that the intersection form Q is even if and only if $c_1(K) \in 2H^2(S, \mathbb{Z})$. Therefore, Freedman's theorem 14.3 has as a corollary that two simply connected complex surfaces S_1, S_2 are homeomorphic if and only if they have the same invariants K^2, χ, and for both of them the same answer holds true to the question: does $K_i \in 2H^1(S_i, \mathbb{Z})$ or does it not?

We note that in complex dimension n = 1, the notion of homeomorphic and diffeomorphic are the same, and that the moduli space $\mathfrak{M}_g$ of curves of genus g is connected, also irreducible, and of pure dimension 3g - 3 (more of its properties are known, cf. [H-M], and Harer's notes in this volume).

In the next paragraphs, we shall consider some families of surfaces, somehow a generalization of hyperelliptic curves, by which we shall see that $\mathfrak{M}^{top}$ does not share any of these three good properties with $\mathfrak{M}_g$. We have in fact the following (cf. [Ca 6], [Ca 8]).

Theorem 21.3. For each natural number k, there exists a minimal surface of general type S such that $\mathfrak{M}^{top}(S)$ has at least k irreducible components

$Y_1, \ldots, Y_k$ with

i) $\dim(Y_i) \neq \dim(Y_j)$ for $i \neq j$

ii) Y_i and Y_j lie, for $i \neq j$, in different connected components of $\mathfrak{M}^{top}(S)$.

It is not clear at the time being how smaller than $\mathfrak{M}^{top}$ is really $\mathfrak{M}^{diff}$: Donaldson [Do 2] has shown the existence of two homeomorphic, but not diffeomorphic, complex surfaces, so we should expect to have $\mathfrak{M}^{diff} \neq \mathfrak{M}^{top}$ in general, and $\mathfrak{M}^{diff}$ could still have some nice properties.

§ 22. Bidouble covers and their deformations

Definition 22.1. A smooth bidouble cover (it is a fourfold cover) is a Galois finite cover, $\varphi: S \to X$ with group $\mathbb{Z}/2 \oplus \mathbb{Z}/2$, between smooth varieties.

Example 22.2. Let $\varphi: \mathbb{P}^2 \to \mathbb{P}^2$ be the morphism defined by $(x_0, x_1, x_2) \to (x_0^2, x_1^2, x_2^2) = (z_0, z_1, z_2)$. The group $(\mathbb{Z}/2)^2$ acts with the three covering involutions $\sigma_0, \sigma_1, \sigma_2$ such that $\sigma_i^*(x_j) = x_j$ if $i \neq j$, $\sigma_i^*(x_i) = -x_i$. The fixed locus for σ_i is the line $\{x_i = 0\} = R_i$ plus the point $R_j \cap R_k$, if (i, j, k) is a permutation of $(1, 2, 3)$, and the branch locus $B = B_1 + B_2 + B_3$ consists of the 3 coordinate lines, respective images of R_1, R_2, R_3. This example, though easy, gives all the ingredients of the geometric picture in the general situation.

As in the example one defines (cf. [Ca 6]) $\sigma_0, \sigma_1, \sigma_2$ to be the three nontrivial elements of the group, and denotes by R_i the divisorial part of the fix locus of σ_i, by B_i the set theoretic image of R_i.

If $z_i = 0$ is a section of $\mathcal{O}_X(B_i)$ with $\operatorname{div}(z_i) = B_i$, we see from the example above that it is not possible to take directly only the square root x_i of z_i, but that it is possible to take the square roots w_i of $z_0 z_1 z_2 / z_i$ (these three double covers correspond to the three distinct subgroups of order 2 of $(\mathbb{Z}/2)^2$, giving each a factorization of φ as a composition of two double covers). In general, thus, there are divisors L_0, L_1, L_2 on X such that $2L_i \equiv B_0 + B_1 + B_2 - B_i$, $B_k + L_k \equiv L_0 + L_1 + L_2 - L_k$, and S is the smooth surface, in the rank 3 bundle on X which is the direct sum of the three line bundles corresponding to the L_i's, defined by the equations

$$(22.3) \qquad \begin{cases} w_i^2 = z_0 z_1 z_2 / z_i \\ z_k w_k = w_0 w_1 w_2 / w_k \end{cases}$$

There is a natural way of deforming these equations, and, computing N_φ and the characteristic map of the family, one can check whether the characteristic system of the morphism φ is complete.

We refer to [Ca 6] for more details, and observe that equations 22.3 take a much simpler form, and the natural way of deforming is easier to see if one assumes one involution, say σ_0, to have only isolated fixed points, i.e., if one assumes $z_0 = 1$.

Definition 22.4. A simple bidouble cover is a smooth bidouble cover such that one of the three covering involutions has a fixed set of codimension at least 2.

The equations 22.3 simplify then (set $x_1 = w_2$, $x_2 = w_1$) to

(22.5)
$$\begin{cases} x_1^2 = z_1 \\ x_2^2 = z_2 , \end{cases}$$

and a natural way to deforming them is to set

(22.6)
$$\begin{cases} x_1^2 = z_1 + b_1 x_2 \\ x_2^2 = z_2 + b_2 x_1 , \end{cases}$$

for b_i a section of $\mathcal{O}_X(B_i - L_i)$. In general, applying φ_* to the exact sequence

$$0 \to \Theta_S \to \varphi^* \Theta_X \to N_\varphi \to 0 ,$$

one obtains

(22.7)
$$0 \to \varphi_* \Theta_S \to \Theta_X \otimes (\mathcal{O}_X \oplus \mathcal{O}_X(-L_1) \oplus \mathcal{O}_X(-L_2) \oplus \mathcal{O}_X(-L_3))$$
$$\to \bigoplus_{i=0}^{2} (\mathcal{O}_{B_i}(B_i) \oplus \mathcal{O}_{B_i}(B_i - L_i)) \to 0$$

Now, the parameter space for the natural deformations of φ is the vector space

$$\bigoplus_{i=0}^{2} H^o(\mathcal{O}_X(B_i) \oplus \mathcal{O}_X(B_i - L_i))$$

and one obtains

Theorem 22.8. The characteristic system of the map φ is complete if $H^o(\mathcal{O}_X(B_i))$ goes onto $H^o(\mathcal{O}_{B_i}(B_i))$, and the same holds for

$$H^o(\mathcal{O}_X(B_i - L_i)) \to H^o(\mathcal{O}_{B_i}(B_i - L_i)) .$$

If furthermore $H^1(\Theta_X) = H^1(\Theta_X(-L_i)) = 0$, then the Kuranishi family of S is smooth.

Actually the hypotheses in the above theorem are rather restrictive, and not strictly needed, anyhow they are sufficient for our application (cf. § 23). Also, a similar result should hold true more generally for smooth Abelian covers.

§ 23. Bihyperelliptic surfaces

Hyperelliptic curves are double covers of $\mathbb{P}^1$, and if we multiply all the previous terms by two we get

Definition 23.1. A bihyperelliptic surface is a smooth bidouble cover of $\mathbb{P}^1 \times \mathbb{P}^1$.

In the following, we shall limit ourselves to consider simple bihyperelliptic surfaces. These are determined by the two branch curves B_1, B_2 of respective bidegrees (2n, 2m), (2a, 2b). One can allow also the two curves B_1, B_2 to acquire singularities, but in such a way that the bidouble cover defined by equations (22.5) have only R.D.P.'s as singularities: we shall call the resulting surfaces admissible.

(23.2) Denote by $\mathcal{N}_{(a,b)(n,m)}$ the subset of the moduli space corresponding to natural deformations of simple bihyperelliptic surfaces with branch loci of bidegrees (2a, 2b), (2n, 2m). Denote further by $\hat{\mathcal{N}}_{(a,b)(n,m)}$ the subset corresponding to admissible surfaces (these are the surfaces whose canonical models are defined by equations (22.6), and occur precisely when those equations give surfaces with at most R.D.P.'s as singularities).

An easy application of theorem 22.8 shows

Theorem 23.3. $\mathcal{N}_{(a,b)(n,m)}$ is a Zariski open irreducible subset of the moduli space. In particular, the closure $\overline{\mathcal{N}}_{(a,b)(n,m)}$ is irreducible (and contains $\hat{\mathcal{N}}_{(a,b)(n,m)}$).

Remark 23.4. i) Clearly

$$\eta_{(a,b)(n,m)} = \eta_{(b,a)(m,n)} = \eta_{(n,m)(a,b)} = \eta_{(m,n)(b,a)} :$$

apart from these trivial equalities, all the $\eta_{(a,b)(n,m)}$'s can be proven to be different (by the inflectionary behaviour of the canonical map of the general surface in the family), and hence they are disjoint by theorem 23.3.

ii) Also, since $\mathbb{P}^1 \times \mathbb{P}^1 = \mathbb{F}_0$ is a deformation of $\mathbb{F}_{2n}$, one can also show (cf. [Ca 1]) that, enlarging the set $\eta_{(a,b)(n,m)}$ to the smooth bidouble covers of $\mathbb{F}_{2n}$, and doing the same for the admissible covers, a result similar to 23.3 holds true.

iii) If $a > 2n$, $m > 2b$, it follows easily from equations (22.6) that all the surfaces in $\hat{\eta}_{(a,b)(n,m)}$ are admissible (simple) bihyperelliptic surfaces.

We can now sketch the main arguments for the proof of theorem 21.3.

(23.5) Bihyperelliptic surfaces are simply-connected, and their invariants K^2, χ are expressed by quadratic polynomials P, Q of (a,b,n,m).

(23.6) Also $\dim \eta_{(a,b)(n,m)}$ is given by an easy function of (a,b,n,m).

(23.7) Letting $r(S) = \max\{m \mid c_1(K) \in mH^2(S,\mathbb{Z})\}$, since for a family $\mathcal{S} \to B$, $\mathcal{O}_{S_t}(K_t) = \mathcal{O}_{S_t} \otimes \omega_{\mathcal{S}|B}$, we have that r is a locally constant function on the moduli space. Moreover (cf. [Ca 7] for the proof, using easy arguments of group cohomology), if $[S] \in \eta_{(a,b)(n,m)}$, then $r(S) = \text{G.C.D.}(a+n-2, b+m-2)$.

(23.8) One has to show (this was done by Bombieri in the appendix to [Ca 6], that for each k, there exist k 4-tuples $(a,b)(n,m)$ giving the same values for K^2, χ, and k distinct values for both $M(S)$ and $r(S)$, and one can further assume $r(S)$ to take even values. But when $w_2 = 0$, Q is even, the S's are simply connected, therefore (cf. 21.2) one gets k distinct irreducible components, of different dimensions, belonging to the same moduli space $\mathfrak{M}^{top}(S)$, and lying in k distinct connected components of $\mathfrak{M}^{top}(S)$. I conjecture the closures of the $\eta_{(a,b)(n,m)}$'s (at least if $a > 2n$, $m > 2b$) to be themselves connected components of the moduli space. The following has been proven up to now ([Ca 7]), and it is an encouraging result, since one of the most difficult problems is, in general, to describe deformations in the large of complex manifolds.

Theorem 23.9. If $a > 2n$, $m > 2b$ the closure of $\eta_{(a,b)(n,m)}$ consists of admissible covers of some $\mathbb{F}_{2k}$, with

$$k \leq \max\left(\frac{n}{a-1}, \frac{n}{m-1}\right).$$

In particular $\hat{\eta}_{(a,b)(n,m)}$ is a closed subvariety of the moduli space if

$$a \geq \max(2n+1, b)$$
$$m \geq \max(2n+1, m).$$

Idea of proof: If $a > 2n$, $m > 2b$, then (cf. 23.4.iii) all the surfaces in $\eta_{(a,b)(n,m)}$ are simple bihyperelliptic and, given a 1-dimensional family $\mathcal{S} \xrightarrow{p'} T$, with $[S_t] \in \eta_{(a,b)(n,m)}$ for $t \neq t_o$, one wants to conclude that S_{t_o} is still an admissible cover of some $\mathbb{F}_{2k}$. The key point is that $(\mathbb{Z}/2)^2$ acts birationally on $\mathcal{S}$, preserving the deformation morphism p', but indeed it acts biregularly on the family $\mathcal{X} \xrightarrow{p} T$ of the canonical models of the previous family $p'\colon \mathcal{S} \to T$. What we have to show is that, if $Z = \mathcal{X}/(\mathbb{Z}/2)^2$, (then $Z_t = \mathbb{P}^1 \times \mathbb{P}^1$ for $t \neq t_o$), then $Z_{t_o} = \mathbb{F}_{2k}$. To achieve this goal it suffices to show that $q\colon Z \to T$ is a smooth fibration, since every deformation of a minimal rational ruled surface with Q even is again a surface of the same type. Now, the singularities that Z_{t_o} can have are quotients of R.D.P.'s by $\mathbb{Z}/2$ or $(\mathbb{Z}/2)^2$, and can be explicitly classified: but many of them can be shown not to occur since any smoothing of these singularities would contribute, through the vanishing cohomology of the Milnor fibre, a subspace of $H^2(Z_t, \mathbb{Z})$ of dimension ≥ 2 over which Q is negative definite. This is a contradiction, since $Z_t \cong \mathbb{P}^1 \times \mathbb{P}^1$, and other arguments again by contradiction, eliminate the other remaining possibilities.

I should finally remark that to prove that the closure of $\eta_{(a,b)(n,m)}$ is open in the moduli space, it would suffice (by the results of [Ca 6] and [Ca 7]) to prove also when the canonical model of S is singular (i.e., the bidouble cover is admissible, but not smooth) that the base B of the Kuranishi family of S is locally irreducible.

References

[Ar] Artin, M., "On the solutions of analytic equations," Invent. Math., 5 (1968), 277-291, MR 38 #344.

[Ar 2] Artin, M., "Algebraization of formal moduli, I, Global Analysis," Princeton Univ. Press, Princeton, N.J., and Tokyo Univ. Press, Tokyo, 1969, 21-71, MR 41#5369.

[A-S] Atiyah, M.I., Singer, I. M.,"The index of elliptic operators III," Ann. Math., 87 (1968), 546-604.

[B-DF] Bagnera, G., De Franchis, M.,"Sopra le superficie algebriche che hanno le coordinate del punto generico esprimibili con funzioni meromorfe quadruplemente periodiche di 2 parametri," Rend. Acc. dei Lincei 16 (1907).

[B-P-V] Barth, W., Peters, C., van de Ven, A., "Compact complex surfaces," Springer Ergebnisse 4, (1984), Berlin-Heidelberg.

[Be 1] Beauville, A., "Surfaces algébriques complexes," Astérisque, 54 (1978).

[Bo] Bombieri, E., "Canonical models of surfaces of general type," Publ. Scient. I.H.E.S. 42 (1973), 447-495.

[Bo-Hu] Bombieri, E., Husemoller, D., "Classification and embeddings of surfaces," in Algebraic Geometry, Arcata 1974, A.M.S. Proc. Symp. Pure Math., 29 (1975), 329-420.

[Brc] Borcea, C., "Some remarks on deformations of Hopf manifolds, Rev. Roum. Math. Pures Appl., 26 (1981), 1287-1294.

[B-W] Burns, D., Wahl, J., "Local contributions to global deformations of surfaces," Inv. Math., 26 (1974), 67-88.

[Cas] Castelnuovo, G., "Sul numero dei moduli di una superficie irregolare, I, II," Rend. Accad. Lincei, 7 (1949), 3-7, 8-11.

[Ca 1] Catanese, F., "Moduli of surfaces of general type," in Algebraic Geometry - Open problems, Proc. Ravello 1982, Springer L.N.M. 997 (1983), 90-112.

[Ca 2] Catanese, F., "Superficie complesse compatte," Atti Convegno G.N.S.A.G.A. del C.N.R., Valetto, Torino (1986)

[Ca 3] Catanese, F., "Canonical rings and "special" surfaces of general type," Proc. of A.M.S. Conf. in Algebraic Geometry, 1985.

[Ca 4] Catanese, F., "Commutative algebra methods and equations of regular surfaces," in Alg. Geometry-Bucharest 1982, Springer L.N.M. 1056 (1984), 68-111.

[Ca 5] Catanese, F., "On the period map of surfaces with $K^2 = \chi = 2$," in Classification of Algebraic and Analytic Manifolds, Proc. Katata Symp. 1982, P.M. 33 Birkhäuser (1983), 27-43.

[Ca 6] Catanese, F., "On the moduli spaces of surfaces of general type," J. Diff. Geom., 19 (1984), 483-515.

[Ca 7] Catanese, F., "Automorphisms of rational double points and moduli spaces of surfaces of general type," Comp. Math. 61 (1987) 81-102.

[Ca 8] Catanese, F., "Connected components of moduli spaces," J. Diff. Geom. 24 (1986) 395-399.

[Ci 1] Ciliberto, C., "Superficie algebriche complesse: idee e metodi della classificazione" in Atti del Convegno di Geometria Algebrica, Nervi 1984, Tecnoprint Bologna (1984), 39-157.

[Ci 2] Ciliberto, C., "Sul grado dei generatori dell' a nello canonico di una superficie di tipo generale," Rend. Sem. Mat. Univ. e Polit. di Torino 41, 3 (1983), 83 - 112.

[Da] Dabrowski, K., "Moduli spaces for Hopf surfaces," Math. Ann., 259 (1982), 201-226.

[Do 1] Donaldson, S., "An application of gauge theory to four dimensional topology," J. Diff. Geom., 18 (1983), 279-315.

[Do 2] Donaldson, S., "La topologie differentielle des surfaces complexes," C.R. Acad. Sc. Paris 301, I, 6 (1985), 317-320.

[Dou 1] Douady, A., "Le probleme des modules pour les variétés analytiques complexes, Sém. Bourbaki (1964/65), exposé 277.

[Dou 2] Douady, A., "Flatness and privilege," in Topics in several complex variables, Monographie n. 17 de l'Ens. Math., Geneve (1968), 47-74.

[Dou 3] Douady, A., "Obstruction primaire à la déformation, Séminaire H. Cartan, 13^c Annee (1960/61), exposé 4.

[Dou 4] Douady, A., "Le problème des modules pour les sous-espaces analytiques compacts d'un espace analytique donné," Ann. Inst. Fourier (Grenoble) 16 (1966), fasc. 1, 1-95, MR 34 #2940.

[Eh] Ehresmann, C., ""Sur les espaces fibrés différentiables," C. R. Acad. Sci. Paris, 224 (1947), 1611-1612.

[En] Enriques, F., "Le superficie algebriche," Zanichelli, Bologna (1949)

[Fo] Forster, O., "Power series methods in deformation theory," Proc. Symp. Pure Math., 30 (1977), A.M.S., 199-217.

[Fra] Francia, P., "The bicanonical map for surfaces of general type," to appear in Math. Ann.

[Fre] Freedman, M. H., "The topology of four dimensional manifolds," J. Diff. Geom., 17 (1982), 357-453.

[Gie] Gieseker, D., "Global moduli for surfaces of general type," Inv. Math. 43 (1977), 233-282.

[Gr 1] Grauert, H., "Über die Deformation isolierter Singularitäten analytischer Mengen," Invent. Math., 15 (1972), 171-182, MR 45 #2206.

[Gr 2] Grauert, H., "Über Modifikationen und exzeptionelle analytische Mengen," Math. Ann., 146 (1962), 331-368.

[Gr 3] Grauert, H., "Der Satz von Kuranishi für kompakte komplexe Raüme, Invent. Math., 25 (1974), 107-142, MR 49 #10920.

[G 1] Griffiths, Ph. A., "Periods of integrals on algebraic manifolds, I. II," Amer. J. Math., 90 (1968), 568-626, 805-865, MR 37 #5215, 38 #2146.

[G-H] Griffiths, P., Harris, J., "Infinitesimal variation of Hodge structure II," Comp. Math 50 (1983), 207-265.

[Gro 1] Grothendieck, A., "Construction de l'espace de Teichmüller, Séminaire H. Cartan, 13^c Année (1960/61), Exposé 17.

[Gro 2] Grothendieck, A., "Techniques de construction II; Le théorème d' existence en théorie formelle des modules, Sém. Bourbaki, exp. 195 Scre. Math. Paris (1960); Techniques de construction IV: Les schémas de Hilbert, Sém. Bourbaki, exp. 221, Secr. Math. Paris (1961).

[Hab] Habegger, N., "Une variété de dimension 4 avec forme d'intersection paire et signature -8," Comm. Math. Helv., 57 (1982), 22-24.

[H-M] Harris, J., Mumford, D., "On the Kodaira dimension of the moduli space of curves," Invent. Math., 67 (1982), 23-86.

[Hi] Hicks, N., "Notes on differential geometry," Van Nostrand, New York (1965).

[Hir] Hirzebruch, F., "Topological methods in algebraic geometry, Springer-Verlag, Grundlehren 131, 3rd ed. (1966).

[Hor 0] Horikawa, E., "On deformations of holomorphic maps," in Manifolds - Tokyo 1973, Iwanami-Shoten (1974), 383-388.

[Hor 1] Horikawa, E., "On deformations of holomorphic maps I," J. Math. Soc. Japan, 25 (1973), 372-396.

[Hor 2] Horikawa, E., "On deformations of holomorphic maps II," J. Math. Soc. Japan 26, 647-667, III ibidem.

[Hor 3] Horikawa, E., "On deformations of quintic surfaces," Inv. Math., 31 (1975), 43-85.

[Hor 4] Horikawa, E., "On the number of moduli of certain algebraic surfaces of general type," J. Fac. Sci. Univ. Tokyo (1974), 67-78.

[Hör] Hörmander, L., "An introduction to complex analysis in several variables," II ed., North Holland, Amsterdam (1973).

[Kas] Kas, A., "On obstructions to deformations of complex analytic surfaces," Proc. Nat. Acad. Sci. U. S. A., 58 (1967), 402-404.

[Ko 1] Kodaira, K., "On stability of compact submanifolds of complex manifolds," Amer. J. Math., 85 (1963), 79-94, MR 27 #3002.

[Ko 2] Kodaira, K., "A theorem of completeness for analytic systems of surfaces, with ordinary singularities," Ann. Math., 74 (1961), 591-627.

[Ko 3] Kodaira, K., "On characteristic systems of families of surfaces with ordinary singularities in a projective space," Amer. J. Math., 87 (1965), 227-256.

[Ko 4] Kodaira, K., "On the structure of compact complex analytic surfaces, I," Amer. J. Math., 86 (1964), 751-798.

[K-M] Kodaira, K., Morrow, J., "Complex manifolds," Holt, Rinehart and Winston, New York (1971).

[K-N-S] Kodaira, K., Nirenberg, L., Spencer, D. C., "On the existence of deformations of complex analytic structures," Ann. Math. (2) 68, (1958), 450-459, MR 22 #3012.

[K-S] Kodaira, K., Spencer, D. C., "On deformations of complex analytic structures, I, II," Ann. Math. (2) 67 (1958), 328-466, MR 22 #3009.

[K-S-2] Kodaira, K., Spencer, D. C., "A theorem of completeness for complex analytic fibre spaces," Acta Math., 100 (1958), 281-294.

[Ku 1] Kuranishi, M., "On the locally complete families of complex analytic structures," Ann. Math. (2), 75 (1962), 536-577, MR 25 #4550.

[Ku2] Kuranishi, M., "New proof for the existence of locally complete families of complex structures, Proc. Conf. Complex Analysis (Minneapolis 1964), Springer-Verlag, Berlin-Heidelberg New York, 1965, 142-154, MR 31 #768.

[Ku 3] Kuranishi, M., "Deformations of compact complex manifolds," Presses de l'Université de Montreal (1969)

[L-S] Lichtenbaum, S., Schlessinger, M., "The cotangent complex of a morphism," Trans. Amer. Math. Soc., 128 (1967), 41-70, MR 25 #4550.

[Man] Mandelbaum, R., "Four dimensional topology: an introduction," Bull. A.M.S. (1980), 1-159.

[Math 1] Mather, J. N., "Stable map germs and algebraic geometry," Manifolds-Amsterdam, 1970, Springer L.N.M. 197 (1971), 176-193.

[Math 2] Mather, J. N., "Generic projections," Annals of Math., 98 (1973), 226-245.

[Mat] Matsumura, H., "On algebraic groups of birational transformations," Rend. Accad. Lincei Ser. 8, 34 (1963), 151-155.

[Mi-Sta] Milnor, J., Stasheff, J., "Characteristic classes," Ann. Math. Studies 76, Princeton Univ. Press, Princeton (1974).

[Moi 1] Moishezon, B. G., "On n-dimensional compact varieties with n algebraically independent meromorphic functions," Izv. Ak Nauk SSSR Ser. Mat. 30 (1966), 133-144, 345-386, 621-656. English transl., Am. Math. Soc. Transl. (2), 63 (1967), 51-177.

[Moi 2] Moishezon, B. G., "An algebraic analogue to compact complex spaces with a sufficiently large field of meromorphic functions. Izv. Akad. Nauk SSSR Ser. Mat., 33 (1969), 174-238, 323-367, 506-548.

[Mu 1] Mumford, D., "Further pathologies in algebraic geometry," Am. J. Math. 84 (1962) ,642-648 .

[Mu 2] Mumford, D., "Geometric invariant theory," 2nd edition (1982), with J. Fogarthy coauthor), Heidelberg, Springer (1965).

[Mu 3] Mumford, D., "Lectures on curves on an algebraic surface," Annals of Math. Studies 59, Princeton Univ. Press, Princeton (1966).

[Nak 1] Nakamura, I., "VII^0 surfaces and a duality of cusp singularities," in Classification of algebraic and analytic manifolds. Proc. Katata Symp. 1982, P.M. 39, Birkhäuser (1983), 333-378.

[Nak 2] Nakamura, I., "On surfaces of class VII_0 with curves," Inv. Math. 78, 3 (1984), 393-443.

[Noe 1] Noether, M., "Zur theorie des eindeutigen Entsprechens algebraischer Gebilde," Math. Ann., 2 (1870), 293-316, 8 (1875), 495-533.

[Noe 2] Noether, M.,"Anzahl der Moduln einer Classe algebraischer Flächen," Sb. Kgl. Preuss, Akad. Wiss. Math.-Nat. Kl., Berlin (1888), 123-127.

[Pa] Palamodov, V. P., "Deformations of complex spaces," Uspekhi Mat. Nauk. 31:3 (1976), 129-194. Transl. Russian Math. Surveys 31:3 (1976), 129-197.

[Pe] Persson, U., "An introduction to the geography of surfaces of general type," Proc. of A.M.S. Conf. in Algebraic Geometry, 1985.

[Po] Popp, H., "Moduli theory and classification theory of algebraic varieties," Springer L.N.M. 620 (1977).

[Pou] Pourcin , G., "Déformation de singularités isolées, Astérisque, Soc. Math. France 16 (1974), 161-173.

[Ram] Ramanujam, C. P., "Remarks on the Kodaira vanishing theorem," J. Indian Math. Soc., 36 (1972), 41-51; supplement in J. Indian Math. Soc. 38 (1974), 121-124.

[Rei 1] Reider, I., "Bounds on the number of moduli for irregular surfaces of general type," Manuscr. Math.

[Rei 2] Reider, I., " Linear systems and vector bundles of rank 2 on surfaces" Annals of Math., to appear.

[Rk] Rokhlin, V. A., "New results in the theory of 4-dimensional manifolds," Dokl. Akad. Nauk. SSSR, 84 (1952), 221-224.

[Ro] Rossi, H., "Vector fields on analytic spaces," Annals of Math. 78, 3 (1963), 455-467.

[Sch] Schlessinger, M., "Functors of Artin rings," Trans. Amer. Math. Soc., 130 (1968), 208-222, MR 36 #184.

[Sei 1] Seiler, W.K., "Global Moduli for Elliptic Sufaces with a Section," Comp. Math. 62 (1987), 169-185.

[Sei 2] Seiler, W.K., "Global Moduli for Polarized Surfaces," Comp. Math. 62 (1987) 187-213.

[Sei 3] Seiler, W.K., "Deformation of Weierstraß Elliptic Surfaces," Math. Annal. 1988

[Ser] Sernesi, E., "Small deformations of global complete intersections," B.U.M.I., 12 (1975), 138-146.

[Ser 2] Sernesi, E., "Sullo schema di Hilbert," preprint.

[Se] Serre, J.-P., "Faisceaux algébriques coherents," Ann. Math., 61 (1955), 197-278.

[Se 2] Serre, J.-P., "Cours d'arithmetique," Presses Universitaires de France, Paris (1970).

[Sh] Shafarevitch, I., et al., "Algebraic surfaces," Proc. Steklov Inst., A.M.S. Translation (1967), Providence.

[Sie 1] Siegel, C. L., "Meromorphe Funktionen auf kompakten analytischen Mannigfaltigkeiten, Nachr. Akad. Wiss. Göttingen, math-phys. Klasse 4 (1955), 71-77.

[Siu] Siu, T. Y., "Every K-3 surface is Kähler," Inv. Math., 73 (1983), 139-150.

[Spiv] Spivak, M., "(A comprehensive introduction to) Differential Geometry," II ed., vol. I-V, Publish or Perish, Berkeley (1979).

[Su] Suwa, T., "Stratification of local moduli spaces of Hirzebruch manifolds," Rice University Studies 59 (1975), 129-146, MR 48 #4362.

[Tju] Tjurina, G. N., "Resolution of singularities of flat deformations of rational doublets," Funktsional. Anal. i Prilozhen., 4 (1) (1970), 77-83, MR 42 #2031.

[To] Todorov, A., "Applications of the Kähler-Einstein-Calabi-Yau metric to moduli of K3-surfaces," Invent. Math., 81 (1980), 251-266.

[Us] Usui, S., "Deformations and local Torelli theorem for certain surfaces of general type," in Algebraic Geometry, Copenhagen 1978, Springer L.N.M. 732 (1979), 605-623.

[Wah] Wahl, S., "Deformations of plane curves with nodes and cusps," Am. J. Math., 96 (1974), 529-577 .

[Wav] Wavrik, J. J., "Obstructions to the existence of a space of moduli," Global Analysis, Princeton Math. Series 29 (1969), 403-414.

[We] Weil, A., "Introduction a l'etude des varietes Kählericnnes," Hermann, Paris (1958).

[Ya] Yau, S. T., "Calabi's conjecture and some new results in algebraic geometry," Proc. Nat. Acad. Sci. U.S.A., 74 (1977), 1798-1799.

[Za] Zariski, O., "Algebraic surfaces, 2nd ed. Erg. Math. 61, Springer, Heidelberg (1971).

[Zp] Zappa, G., "Sui Sistemi continui di curve sopra una rigata algebrica," Gior. Mat. Battaglini, 77 (1947).

[Zp 2] Zappa, G., "Sulla degenerazione delle superficie algebriche in sistemi di piani distinti, con applicazioni allo studio de llerigate," Atti Accad. Naz. Lincei Mem. ,13 (1942) .

THE SCHOTTKY PROBLEM

Ron Donagi
Department of Mathematics
University of Pennsylvania
Philadelphia, PA 19104
USA

Introduction

The Schottky problem is the problem of characterizing the Jacobians of algebraic curves among all principally polarized abelian varieties. The three most successful approaches to this problem are associated with the names of Schottky-Jung, Andreotti-Mayer, and Novikov. The special properties used to characterize Jacobians are, respectively, the existence of Prym varieties, the large singular locus of a Jacobian theta divisor, and the differential equations satisfied by Jacobian theta functions.

The purpose of these notes is to describe how closely each approach is known to characterize Jacobians, and especially to relate these three apriori independent approaches.

Novikov's approach has come the closest. The Novikov Conjecture, proved by Shiota, says that an <u>indecomposable</u> abelian variety X is a Jacobian if there are 3 translation-invariant vector fields D_1, D_2, D_3 on X and a scalar d so that the theta function of X satisfies the "KP" differential equation corresponding to these constants. In some ways, though, this solution is still not satisfactory. First, it is not clear how to convert the KP equation into explicit equations in the natural coordinates on moduli space, the theta nulls. Then there is the question of eliminating the choices (D_1, D_2, D_3, d) involved, and the fact that the locus we get in $\mathcal{A}_g$ consists of Jacobians $\mathcal{J}_g$ together with all decomposable abelian varieties. Finally, from an algebro-geometric point of view, Novikov's condition may be considered to be "too strong". We discuss the differential-equations approach in Chapter 3, and in particular we propose a stronger version (3.1) of the Novikov Conjecture, i.e. a weaker condition which should suffice to characterize Jacobians among indecomposable abelian varieties.

The Andreotti-Mayer condition, on the other hand, is too weak. The locus defined by it contains Jacobians as an irreducible component, but does contain other components. Beauville and Debarre have shown that the Andreotti-Mayer locus contains Novikov's locus as well as several of its variants. We discuss this approach in general in Chapter 1, then return to it in Chapter 4 with an analysis of the components of the Andreotti-Mayer locus in genus ≤ 5.

The strength of the Schottky-Jung approach is somewhere in between. The conjecture is much stronger (i.e. the condition satisfied by Jacobians is much weaker) than Novikov's, yet I am aware of no evidence against it. A precise version of the conjecture is stated in (2.11). It implies the strong version of Novikov (3.1), as well as the four conjectures (2.13-2.16) of van Geemen and van der Geer, and various analogues. Again, we first discuss the general theory, including a proof that Jacobians are a component of Schottky, in Chapter 2, and then in Chapter 5 we sketch proofs of Conjecture (2.11) in genus 4 (Igusa's theorem) and genus 5.

These notes evolved from lecture series which I gave at the CIME meeting in Montecatini and at UNAM in Mexico City, in the spring and summer of 1985. The lectures included background material on moduli spaces and their compacitifications (Satake-Baily-Borel, toroidal, stable curves), the algebraic theory of theta functions, the theory of Prym varieties, and the three approaches to the Schottky problem. The original version of these notes has become much too long for the present format; I hope it will appear in the near future as a book on the moduli of curves and abelian varieties. The present version is more or less an extract from the last chapters of the book.

I heartily thank Eduardo Sernesi and Felix Recillas for the invitations to give the lectures which started this project, and for the warmth of their hospitality. I am also grateful to the many people with whom I discussed the Schottky problem over the years, including Arbarello, Beauville, Clemens, Debarre, Mumford, van Geemen, van der Geer and many others. special thanks go to the Max-Planck-Institut für Mathematik which provided the perfect conditions for completing this work, and to Karin Deutler for the excellent typing job.

CHAPTER 0
Theta functions

We recall some notation and results from the theory of theta functions. The reader may prefer to skim through this introductory chapter, or to skip it and refer to it later as needed.

The moduli space $\mathcal{A}_g$ of g-dimensional principally polarized abelian varieties (ppav) is the quotient

$$\mathcal{A}_g := \mathbb{H}_g/\Gamma_g^{(1)}$$

where $\mathbb{H}_g$ is Siegel's half space

$$\mathbb{H}_g := \{\Omega \text{ symmetric } g\times g \text{ complex matrix, } \mathrm{im}(\Omega)>0\},$$

and

$$\Gamma_g^{(1)} := Sp(2g,\mathbb{Z})$$

is the integral symplectic group, acting on $\mathbb{H}_g$ properly discontinuously. For a subgroup

$$\Gamma_g^{\text{level}} \subset \Gamma_g^{(1)}$$

of finite index we have the corresponding level moduli space

$$\mathcal{A}_g^{\text{level}} := \mathbb{H}_g/\Gamma_g^{\text{level}},$$

a finite branched cover of $\mathcal{A}_g$.

The action of Γ_g^{level} on $\mathbb{H}_g$ lifts to a properly discontinuous action of $\Gamma_g^{\text{level}} \ltimes \mathbb{Z}^{2g}$ on $\mathbb{H}_g \times \mathbb{C}^g$, where

$$\gamma = \begin{pmatrix} A & B \\ C & D \end{pmatrix} \in \Gamma_g^{\text{level}}$$

acts on:

$$
\text{(0.1)} \quad
\begin{array}{lll}
\cdot\ \mathbb{H}_g & \text{by} & \Omega \longmapsto (A\Omega + B)(C\Omega + D)^{-1} \\
\cdot\ \mathbb{C}^g & \text{by} & {}^t(C\Omega + D)^{-1} \\
\cdot\ \mathbb{Z}^{2g} & \text{by} & {}^t\gamma^{-1} = \begin{pmatrix} D & -C \\ -B & A \end{pmatrix}.
\end{array}
$$

We let $\mathfrak{X}_g$ denote the quotient. It maps naturally to $\mathcal{A}_g^{\text{level}}$, and for high enough level (i.e. small enough Γ_g^{level} in Γ_g) it gives a universal abelian variety, i.e. the fiber over the isomorphism class [X] is isomorphic to X. On the other hand, if $(-1) \in \Gamma_g^{\text{level}}$ then the generic fiber is the Kummer variety

$$\text{(0.2)} \qquad K(X) := X/(\pm 1).$$

The level groups which we will encounter are:

(0.3) The principal congruence subgroup

$$\Gamma^{(n)} := \langle \gamma \in \Gamma^{(1)} \mid \gamma \equiv 1 \text{ mod. } n \rangle.$$

The quotient $\mathcal{A}_g^{(n)}$ parametrizes ppav's A with a chosen basis for the finite group A_n of points of order n in A. For $n \geq 3$, the action of $\Gamma^{(n)}$ is fixed point free (so $\mathcal{A}_g^{(n)}$ is non-singular) and $\mathfrak{X}_g^{(n)}$ is a universal abelian variety. For $n = 1,2$, $\mathcal{A}_g^{(n)}$ has quotient singularities and the generic fiber of $\mathfrak{X}_g^{(n)} \longrightarrow \mathcal{A}_g^{(n)}$ is a Kummer

(0.4) The theta group

$$\Gamma_g^{(2,4)} := \langle \gamma = \begin{pmatrix} A & B \\ C & D \end{pmatrix} \in \Gamma_g^{(2)} \mid \text{diag}\ {}^tAC \equiv \text{diag}\ {}^tBD \equiv 1 \text{ mod. } 4 \rangle.$$

This sits between $\Gamma_g^{(2)}$ and $\Gamma_g^{(4)}$. The universal object $\mathfrak{X}_g^{(2,4)}$ is a family of Kummer varieties.

(0.5) Fix a primitive vector $v \in \mathbb{Z}^{2g}$, and let

$$\mathfrak{R}\Gamma_g := \{\gamma \in \Gamma \mid \gamma v \equiv v \text{ mod. } 2\},$$

$$\mathfrak{R}\Gamma_g^{(2,4)} := \{\gamma \in \Gamma^{(2,4)} \mid \gamma v \equiv v \text{ mod. } 4\}.$$

The corresponding moduli spaces will be denoted $\mathfrak{R}\mathcal{A}_g$, $\mathfrak{R}\mathcal{A}_g^{(2,4)}$. Here $\mathfrak{R}\mathcal{A}_g$ parametrizes isomorphism classes of ppav's $A \in \mathcal{A}_g$ with a marked, non-zero point of order 2, $\mu \in A_2$, and $\mathfrak{R}\mathcal{A}_g^{(2,4)}$ has a similar interpretation with a level (2,4)-structure and a point of order 4.

The natural coordinates on these spaces are given by the various types of theta function. Riemann's theta function is

(0.6)
$$\vartheta : \mathbb{H}_g \times \mathbb{C}^g \longrightarrow \mathbb{C}$$

$$\vartheta(\Omega, z) := \sum_{n \in \mathbb{Z}^g} \exp \pi i({}^t n \Omega n + 2\,{}^t n z).$$

For given Ω, it is a section of a line bundle $L = \mathcal{O}_X(\Theta)$ on the ppav $X = X_\Omega$. L is in the principal polarization on X; taking a different Ω with the same X may cause L to be changed via translation by a point of order 2 in X. The line bundle L^2 therefore depends on $X \in \mathcal{A}_g$ alone.

The theta divisor $\Theta \subset X$ is given by the vanishing of ϑ. It is a "theta characteristic", i.e. a symmetric divisor representing the principal polarization. Its translates by points of order 2 in X are the other theta characteristics.

For $\epsilon, \delta \in \mathbb{Q}^g$, we have the "theta functions with characteristics":

(0.7)
$$\vartheta[{\epsilon \atop \delta}](\Omega, z) := \sum_{n \in \mathbb{Z}^g} \exp \pi i({}^t(n+\epsilon)\Omega(n+\epsilon) + 2\,{}^t(n+\epsilon)(z+\delta))$$

$$= \exp \pi i({}^t\epsilon\Omega\epsilon + 2\,{}^t\epsilon(z+\delta)) \cdot \vartheta(\Omega, z+\Omega\epsilon+\delta).$$

The space of sections $H^0(X_\Omega, L^k)$ is k^g-dimensional, and an explicit basis is given by the "k-th order theta functions":

$$(0.8) \qquad \vartheta_k[\epsilon](\Omega,z) := \vartheta[{}^{\epsilon/k}_{0}](k\Omega,kz)$$

for $\epsilon \in (\mathbb{Z}/k\mathbb{Z})^g$ (they depend on ϵ only modulo k).

We will make much use of the map given by the second order theta functions

$$(0.9) \qquad \vartheta_2 : \mathbb{H}_g \times \mathbb{C}^g \longrightarrow U_g$$

$$\vartheta_2(\Omega,z)_\epsilon := \vartheta_2[\epsilon](\Omega,z)$$

(where U_g is the standard representation of the Heisenberg group, or we can think of it as shorthand notation for $\mathbb{C}^{2^g}$.) We also consider the projectivized version of ϑ_2:

$$\chi : \mathbb{H}_g \times \mathbb{C}^g \longrightarrow \mathbb{P}(U_g),$$

called the Kummer map. From the transformation properties of ϑ it follows that χ factors through the universal Kummer variety:

$$(0.10) \qquad \chi : \mathfrak{X}_g^{(2,4)} \longrightarrow \mathbb{P}(U_g).$$

Two other theta maps which we will need are

$$\alpha : \mathcal{A}_g^{(2,4)} \longrightarrow \mathbb{P}(U_g)$$

and

$$\beta : \mathcal{R}\mathcal{A}_g^{(2,4)} \longrightarrow \mathbb{P}(U_{g-1})$$

obtained by restricting χ to the 0-section of

$$\mathcal{X}_g^{(2,4)} \longrightarrow \mathcal{A}_g^{(2,4)},$$

respectively to the natural section (which is torsion of order 4) of

$$\mathcal{RX}_g^{(2,4)} \longrightarrow \mathcal{RA}_g^{(2,4)}.$$

At level-∞, these maps are given explicitly by:

$$(0.11)\qquad \begin{aligned} \alpha(\Omega)_\epsilon &:= \vartheta_2[\epsilon](\Omega,0) = \vartheta\begin{bmatrix}\epsilon/2\\ 0\end{bmatrix}(2\Omega,0) \\ \beta(\Omega)_\epsilon &:= \vartheta\begin{bmatrix}\epsilon/2 & 0\\ 0 & 1/2\end{bmatrix}(2\Omega,0). \end{aligned}$$

The entries of α are called the theta nulls. Finally, we can get rid of the annoying level on the left, by dividing by the Galois group

$$G_g(2,4) := \Gamma_g^{(1)}/\Gamma_g^{(2,4)}$$

of $\mathcal{A}_g^{(2,4)}$ over $\mathcal{A}_g$, on the right. We get maps:

$$(0.12)\qquad \begin{aligned} \alpha &: \mathcal{A}_g \longrightarrow \mathbb{P}(U_g)/G_g(2,4) \\ \beta &: \mathcal{RA}_g \longrightarrow \mathbb{P}(U_{g-1})/G_{g-1}(2,4). \end{aligned}$$

($G_g(2,4)$ has a natural, "Heisenberg" action on $\mathbb{P}(U_g)$. We omit the details.)

There is one basic identity relating ϑ to the second-order theta functions:

(0.13) <u>Riemann's Quadratic Identity:</u>

$$\vartheta(\Omega,z+w)\cdot\vartheta(\Omega,z-w) = \sum_{\sigma\in(\mathbb{Z}/2\mathbb{Z})^g} \vartheta_2[\sigma](\Omega,z)\cdot\vartheta_2[\sigma](\Omega,w).$$

Geometrically, this gives an identification of the Kummer map χ with a "dual" map χ^*. For fixed X, χ is simply the map given by the linear system $|\mathcal{O}_X(2\Theta)| = |L^2|$ on X:

$$\chi : X \longrightarrow \mathbb{P}H^0(X,\mathcal{O}_X(2\Theta))^*.$$

On the other hand, fix a theta characteristic Θ, e.g. by choosing a period matrix Ω for X and taking Θ as $\{z|\vartheta(\Omega,z) = 0\}$. We get a map

$$\chi^* : X \longrightarrow \mathbb{P}H^0(X,\mathcal{O}_X(2\Theta))$$

$$x \longmapsto \Theta_x + \Theta_{-x}$$

where Θ_x means the translate of Θ by x. Riemann's Quadratic Identity can be reformulated as:

(0.14) <u>Kummer Identification Theorem</u> There is a (natural) isomorphism $\mathbb{P}H^0(X,\mathcal{O}_X(2\Theta)) \xrightarrow{\sim} \mathbb{P}H^0(X,\mathcal{O}_X(2\Theta))^*$ which takes χ to χ^*.

We mention one more property of the theta function. It satisfies the following analogue of the heat equation:

$$\frac{\partial\vartheta}{\partial\Omega_{ij}} = \frac{1}{2\pi i(1+\delta_{ij})}\,\frac{\partial^2\vartheta}{\partial z_i\partial z_j}\,. \tag{0.15}$$

More algebraically, this gives a natural identification of the tangent space to moduli,

$$T_X\mathcal{A}_g\quad,$$

with the dual of the "quadratic differentials"

$$S^2T_0X.$$

CHAPTER 1
Andreotti-Mayer

The approach based on the singularities of θ is the least successful of the three approaches which we consider. Historically it was the first, and conceptually it is the easiest, so we start with it. We define the Andreotti-Mayer locus and sketch the proof that Jacobians form a component. We also touch on some rather deep results of Green and Welters which are closely related to the Andreotti-Mayer approach but also reappear elsewhere in these notes.

§ 1.1 Sing(θ) for Jacobians

The basic property of Jacobians used in the approach of Andreotti-Mayer is:

(1.1.) Proposition. Let θ be the theta divisor of a Jacobian $J(C)$ of a curve $C \in \mathcal{M}_g$. Then:

$$\operatorname{Dim}(\operatorname{Sing}(\theta)) \geq g - 4.$$

This is based on Riemann's Theorem and Riemann's Singularity Theorem, which say that θ, Sing(θ) on a Jacobian can be described in terms of linear systems on the underlying curve:

$$\theta \approx W^0_{g-1}$$

$$\operatorname{Sing}(\theta) \approx W^1_{g-1}$$

where W^r_d is the subvariety of $\operatorname{Pic}^d(C) \approx J(C)$ consisting of line bundles of degree d on C with $h^0 \geq r + 1$. The well-developed theory of linear systems on curves [ACGH] provides many ways to estimate the dimensions of these varieties:

(1) This is a special case of the Existence Theorem in Brill-Noether theroy, since for

$$r = 1, \qquad d = g - 1$$

we have

$$\rho := g - (r+1)(g-d+r) = g - 2\cdot 2 = g - 4.$$

(2) An elementary argument for this special case is based on checking that for any divisor

$$D_0 = p_1 + \ldots + p_{g-3} \in S^{g-3}C$$

there is a divisor

$$D = D_0 + p_{g-2} + p_{g-1} \in S^{g-1}C$$

such that $h^0(D) \geq 2$. Generically the inequality will be an equality, so we lose only 1 dimension in mapping to $\mathrm{Pic}(C)$.

(3) We give another argument which introduces the very important relationship of $\mathrm{Sing}(\theta)$ with quadrics of rank 4. Assume we are given

$$\xi \in W^1_{g-1} \setminus W^2_{g-1},$$

i.e. a double point of θ. The projectivized tangent cone $\mathbb{P}T_\xi(\theta)$ is thus a quadric Q. By Riemann's Singularity Theorem,

$$Q = \bigcup_{D \in |\xi|} \mathrm{span}(\phi(D))$$

where ϕ is the canonical map of C. Q therefore contains a 1-parameter family of linear subspaces of codimension 1 in Q, or of codimension 2 in $\mathbb{P}^{g-1}$. Therefore Q is a quadric of rank ≤ 4. Its vertex is therefore a linear subspace of codimension ≤ 4 which is contained in $T_\xi(\mathrm{Sing}(\theta))$.

§ 1.2. The Andreotti-Mayer loci

We want to define a series of loci

$$\mathcal{N}_g^k \subset \mathcal{A}_g$$

given by the property:

$$\text{"}A \in \mathcal{N}_g^k \quad \text{if} \quad \dim(\mathrm{Sing}(\theta_A)) \geq k\text{".}$$

To avoid problems of existence and smoothness of the universal abelian variety $\mathfrak{X}_g$, we work initially at level ∞, i.e. in $\mathbb{H}_g$. We define:

$$\mathcal{B}_g := \{(\Omega,z) \in \mathfrak{X}_g^{(\infty)} = \mathbb{H}_g \times \mathbb{C}^g/\mathbb{Z}^{2g} \mid z \in \mathrm{Sing}(\theta_\Omega)\}$$

$\pi : \mathcal{B}_g \longrightarrow \mathcal{A}_g^{(\infty)} = \mathbb{H}_g$, the natural projection.

$$\mathcal{B}_g^k := \{(\Omega,z) \in \mathcal{B}_g \mid \dim_{(\Omega,z)} \pi^{-1}(\Omega) \geq k\} \qquad k = 0,1,\ldots$$

$$\mathcal{N}_g^k := \pi(\mathcal{B}_g^k) \subset \mathcal{A}_g^{(\infty)} = \mathbb{H}_g$$

$\pi^k : \mathcal{B}_g^k \longrightarrow \mathcal{N}_g^k$, the restriction of π.

By construction, $\mathcal{N}_g^k$ is in $\mathbb{H}_g$, but it is clearly $\Gamma^{(1)}$-invariant, so it determines a locus in $\mathcal{A}_g$ which we also denote $\mathcal{N}_g^k$. The Andreotti-Mayer locus is then [AM]:

$$\mathcal{AM}_g := \mathcal{N}_g^{g-4} \subset \mathcal{A}_g.$$

§ 1.3 Jacobians are a component of Andreotti-Mayer

(1.2) Theorem [AM] $\mathcal{J}_g$ is an irreducible component of $\mathcal{AM}_g$.

Proposition (1.1) tells us that $\mathcal{J}_g \subset \mathcal{AM}_g$. The idea is to show that at a generic $C \in \mathcal{M}_g$ we have an equality of tangent spaces

$$T_{J(C)}\mathcal{J}_g \approx T_{J(C)}\mathcal{AM}_g,$$

or equivalently that the conormal spaces agree. The heat equation gives an interpretation of quadrics in canonical space as cotangent directions, at $J(C)$, to $\mathcal{A}_g$. With this interpretation, the conormal to $\mathcal{J}_g$ becomes the space

$$I_2 := \ker(S^2H^0(\omega_C) \longrightarrow H^0(\omega_C^{\otimes 2}))$$

of quadrics through the canonical curve. We claim that the conormal to $\mathcal{AM}_g$ is given by

$$I_2(\theta) := \operatorname{span}\langle (\frac{\partial^2 \vartheta}{\partial z_i \partial z_j})\Big|_{\xi} \;\; | \xi \in \operatorname{Sing}(\theta)\rangle.$$

Note that $I_2(\theta)$ is a subspace of I_2, by Riemann's Singularity Theorem.

(1.3) <u>Lemma</u> Let

$\overline{X}$ be a curve in $\mathcal{B}_g^0$,
$X := \pi(\overline{X}) \subset \mathcal{N}_g^0$, its projection,
$(x,\xi) \in \overline{X}$, a point.

Then the tangent cone $T_xX \subset T_x\mathbb{H}_g$ to the curve X at x is contained in the hyperplane

$$(\frac{\partial \vartheta}{\partial \Omega})\Big|_{(x,\xi)} = 0.$$

(This lemma follows immediately when we differentiate ϑ along X, using the vanishing

$$\vartheta = \frac{\partial \vartheta}{\partial z} = 0.)$$

We can conclude that $I_2(\vartheta)$ is <u>contained</u> in the conormal to $\mathcal{AM}_g$ at $J(C)$, by combining the lemma with the heat equation (0.15) and with some sort of irreducibility assumption, for instance it suffices to assume:

(A1) $\mathrm{Sing}(\theta)$ is precisely $(g-4)$-dimensional.
(A2) $\mathrm{Sing}(\theta)$ is irreducible.

(These assumptions imply that for any curve $X \subset \mathcal{AM}_g$ passing through the point $J(C) \in \mathcal{J}_g$, and any $\xi \in \mathrm{Sing}(\theta_C)$, there is a lift $X_\xi \subset \mathcal{B}_g^{g-4}$ of X passing through $(J(C),\xi)$, so we can apply (1.3) to X_ξ.)

To prove Theorem (1.2) it therefore suffices to exhibit a curve X satisfying (A1), (A2) and:

(A3) $I_2 = I_2(\theta)$.

The argument clearly breaks down without the irreduciblity (A2), since quadrics coming from points ξ in different components of $\mathrm{Sing}(\theta)$ could give directions normal to different curves X in $\mathcal{AM}_g$. Still, we may weaken (A2) to:

(A2') For each component Ξ of $\mathrm{Sing}(\theta)$, the quadrics

$$\left\{\left(\frac{\partial^2\vartheta}{\partial z_i \partial z_j}\right)\Big|_\xi \;\Big|\; \xi \in \Xi\right\}$$

span $I_2(\theta)$.

In the original proof [AM], Andreotti and Mayer consider trigonal curves C. Here (A2) fails, but (A1) and (A2') are easy: $\mathrm{Sing}(\theta)$ consists of two components, each $(g-4)$-dimensional. One is

$$\{L = L_0 \otimes T \mid L_0 \in W^0_{g-4},\ T = \text{the trigonal bundle}\},$$

the other is its image under the involution

$$L \longmapsto \omega_C \otimes L^{-1}.$$

Symmetry implies that the components span the same subspace of $I_2(\theta)$, which is therefore all of $I_2(\theta)$. This explicit description of $\mathrm{Sing}(\theta)$ then allows direct verification of (A3), proving the theorem.

QED

§ 1.4 Further results

It turns out that all three assumptions made in the proof of Theorem (1.2) hold, at least generically. We discuss these next.

(A1) By (1.1), we know that $\dim(\mathrm{Sing}(\theta))$ is always at least $g - 4$. An easy dimension count shows that equality must hold generically. A theorem of Martens [ACGH, p. 191] says that equality holds if and only if C is non-hyperelliptic.

(A2) The irreducibility of $\mathrm{Sing}(\theta)$ for generic C also follows from Brill-Noether theory. In fact, the Fulton-Lazarsfeld Connectedness Theorem together with Gieseker's Smoothness Theorem [ACGH, pp. 212 and 214] imply that W^r_d is irreducible for generic C whenever the Brill-Noether number ρ is ≥ 1.

A more precise result is known in our case, when $r = 1$: Teixidor [Tx] shows that W^1_d is irreducible except when C is trigonal, bielliptic (branched double cover of an elliptic curve) or a certain type of curve of genus 5.

(A3) Andreotti and Mayer showed that $I_2(\theta) = I_2$ for trigonal C, hence for generic C. There are several other loci where the equality can be checked directly, e.g. for bielliptic curves. The best result was proved by Mark Green:

(1.4) <u>Theorem [G]</u> For any non-hyperelliptic curve C of genus $g \geq 4$, the space I_2 of quadrics through the canonical curve $\phi(C)$ is spanned by the tangent cones to θ_C at its double points, i.e.

$$I_2(\theta) = I_2.$$

In particular, this implies that I_2 is spanned by quadrics of rank ≤ 4, since we saw in the third proof of (1.1) that tangent cones to θ at double points are quadrics of rank ≤ 4. This also produces a simple proof of Torelli's Theorem, in fact a recipe for recovering a curve (not hyperelliptic, trigonal or a plane quintic) from its Jacobian: the canonical curve $\phi(C)$ is the intersection of the tangent cones to θ at its double points.

Next we descirbe a result of Welters' which is closely related to Green's Theorem. Given a curve C, we define three loci in $J(C)$:

$$F_C := \bigcap_{D\in|2\theta|, m_0(D)\geq 4} (D),$$

$$F'_C := \bigcap_{\xi\in \mathrm{Sing}(\theta)} (\theta_\xi \cup \theta_{-\xi}),$$

$$F''_C := \bigcap_{\xi\in \mathrm{Sing}(\theta)} (\theta_\xi) = \{a \in J(C) \mid a + \mathrm{sing}(\theta) \subset \theta\}.$$

(1.5) Theorem (Welters [We]) For a curve C of genus g, the surface $(C - C) \subset J(C)$ is equal to:

(1) F_C, if $g = 3$ or $g \geq 5$.
(2) F'_C, if $g \geq 5$ and C is not trigonal
(3) F''_C, if $g \geq 5$.

When C is of genus 4, it has two trigonal bundles T_0, T_1 (possibly equal); in this case

$$F_C = (C - C) \cup \pm(T_0 - T_1).$$

We observe that for non-trigonal C, (2) $\Rightarrow$ (1), since

$$(\theta_\xi \cup \theta_{-\xi}) \in \{D \in |2\theta|, m_0(D) \geq 4\}.$$

By Teixidor's results [Tx] on the irreduciblity of $\mathrm{Sing}(\theta)$, we can also deduce (3) $\Rightarrow$ (2), so the main difficulty is in proving (3).

One connection with Green's Theorem is given by the following weak version of (1.5), which follows from (1.4):

(1.6) Corollary (Weak vesion of Welters' Theorem) $C - C$ is a component of F_C $(g \geq 4)$ and F'_C, F''_C $(g \geq 5)$.

Since clearly $C - C$ is contained in the three loci, it suffices to show that they are 2-dimensional at 0. Green's Theorem says that

$$C = \cap \mathbb{P}T_\xi\theta,$$

hence set-theoretically

$$C = \mathbb{P}T_0(\cap_\xi \ (\theta_\xi \cup \theta_{-\xi}))$$

so the tangent cone is the cone over C, a surface (and of course, equal exactly to $T_0(C - C)$).

The main connection of the two theorems is in the proofs, both of which make heavy use of the geometry of the $(g - 1)$-st symmetric product $S^{g-1}C$, which is a desingularization of θ.

CHAPTER 2
Schottky-Jung

After reviewing some basic properties of Prym varieties, we define the Schottky loci (there are several of them: $\mathcal{RS}_g$, $\mathcal{S}_g^{(big)}$, $\mathcal{S}_g$) in §2.2, and show that Jacobians are in these loci. The main fact known about these loci is that Jacobians are actually a component; we sketch that in §2.3, and then conclude with a series of conjectures, all of which follow from what should be considered "The Schottky-Jung Conjecture", (2.11).

§ 2.1 Prym varieties

The property of Jacobians used in the Schottky-Jung approach is the existence of Prym varieties. In this section we briefly review the definition and some basic facts about Pryms.

Consider an unramified double cover

$$\pi : \tilde{C} \longrightarrow C$$

of a curve $C \in \mathcal{M}_g$. By Hurwitz' formula, the genus of $\tilde{C}$ is $2g - 1$. For given C, the set of double covers π is in $1 - 1$ correspondence with the set

$$J_2(C) \setminus (0)$$

of nonzero points μ or order 2 in $J(C)$. There are induced maps on Jacobians,

$$\pi^* : J(C) \longrightarrow J(\tilde{C})$$

$$\mathrm{Nm} : J(\tilde{C}) \longrightarrow J(C).$$

The kernel of π^* is $(0,\mu)$, where $\mu \in J_2(C)$ corresponds to π as above. The kernel of Nm also has two components which we denote

P, P^-, where $P \subset J(\tilde{C})$ is an abelian subvariety, and P^- a translate of P by a point of order 2. Since Nm is surjective, P is $(g-1)$-dimensional. The principal polarization on $J(\tilde{C})$ induces twice a principal polarization on P; more precisely:

(2.1) <u>Wirtinger's Theorem [M1]</u> Riemann's theta divisor $\tilde{\Theta} \subset J(\tilde{C})$ intersects P in twice a divisor Ξ in the principal polarization:

$$\tilde{\Theta} \cap P = 2 \cdot \Xi.$$

In particular, we can think of P in a natural way as a ppav, $(P,\Xi) \in \mathcal{A}_{g-1}$, called the Prym variety of (C,μ). The assignment

$$(C,\mu) \longmapsto P = P(C,\mu)$$

gives a morphism of moduli spaces

$$\mathcal{P} : \mathcal{RM}_g \longrightarrow \mathcal{A}_{g-1}$$

called the Prym map.

Let $J, \tilde{J}$ denote $J(C)$, $J(\tilde{C})$ respectively, and let J', $\tilde{J}'$ denote the respective torsers (= principal homogeneous spaces) of effective divisors in the principal polarizations of $J, \tilde{J}$. We have a pullback map

$$\pi^{*\prime} : J' \longrightarrow \tilde{J}'$$

and a pushforward

$$\pi'_* : \tilde{J}' \longrightarrow \{\text{divisors in twice the principal polarization}\}.$$

(2.2) <u>Splitting Theorem</u> For any divisor $\Theta \in J'$ in the principal polarization on J, the pushforward of $\pi^{*\prime}\Theta$ splits:

$$\pi'_*(\pi^{*\prime}(\Theta)) = \Theta + \Theta_\mu .$$

(2.3) <u>Prym-Kummer Identification Theorem [M2]</u> Let L_0 be a line bundle on C satisfying $L_0^{\otimes 2} \approx \mu \otimes \omega_C$, and let $L := \pi^* L_0$. (We think of L_0, L as elements of J', $\tilde{J}'$ respectivley.) Then:

(1) L determines a subvariety $P_L \subset \tilde{J}'$, a translate of P.
(2) L_0 determines a natural (i.e. equivariant under the action of the Heisenberg group) embedding

$$i_{L_0} : |2\Xi|^* \longrightarrow |2\theta|$$

(where Ξ, θ are the natural theta divisors on P, J).

(3) The Kummer map χ_P can be identified with π'_*, i.e. the following diagram commutes:

$$\begin{array}{ccc} P & \xrightarrow{\ \chi_P\ } & |2\Xi|^* \\ \Big\downarrow \int & & \Big\downarrow i_{L_0} \\ P_L & \xrightarrow{\ \pi'_*\ } & |2\theta|. \end{array}$$

We also mention that the Abel-Jacobi map

$$AJ : \tilde{C} \longrightarrow \tilde{J}$$

induces an "Abel-Prym" map

$$AP : \tilde{C} \longrightarrow P.$$

While the "derivative" (Gauss map) of AJ is the canonical map, the derivative of AP is the Prym-canonical map $C \longrightarrow \mathbb{P}^{g-2}$ given by the linear system $\omega_C \otimes \mu$.

§ 2.2. The Schottky loci

At level-∞, we define the Schottky locus to be:

$$\mathcal{S}_g^{(\infty)} := \beta^{-1}(\text{image } \alpha) \subset \mathbb{H}_g.$$

From the transformation properties of theta functions it follows that $\mathcal{S}_g^{(\infty)}$ is the inverse of a locus $\mathcal{RS}_g \subset \mathcal{RA}_g$, defined by

$$\mathcal{RS}_g := \beta^{-1}(\text{image } \alpha)$$
$$= \left\{ (X,\mu) \in \mathcal{RA}_g \,\middle|\, \begin{array}{l} \chi_X(\tfrac{1}{2}\mu) = \chi_P(0) \quad \text{for some choice} \\ \text{of } \tfrac{1}{2}\mu \in X \quad \text{and some} \quad P \in \mathcal{A}_{g-1} \end{array} \right\}.$$

(The last condition can also be interpreted as an equality of sets in $\mathbb{P}(U_{g-1})/G_{g-1}$:

$$\chi_X(\tfrac{1}{2}\mu) = \chi_P(P_2)$$

with $\frac{1}{2}\mu := \{\lambda \in X_4 \mid 2\lambda = \mu\}$.) However, $\mathcal{S}_g^{(\infty)}$ does not come from a locus in $\mathcal{A}_g$ (equivalently, $\mathcal{S}_g^{(\infty)}$ is $\mathcal{R}\Gamma$-invariant but not necessarily $\Gamma^{(1)}$-invariant). We therefore have two loci in $\mathcal{A}_g$:

$$\mathcal{S}_g^{(\text{big})} := \{X \in \mathcal{A}_g \mid (X,\mu) \in \mathcal{RA}_g \quad \text{for some} \quad \mu \in X_2 \setminus 0\}$$

$$\mathcal{S}_g := \{X \in \mathcal{A}_g \mid (X,\mu) \in \mathcal{RA}_g \quad \text{for all} \quad \mu \in X_2 \setminus 0\}.$$

(2.4) Schottky-Jung Theorem ([S],[SJ],[FR],[F],[M2]). $\mathcal{RJ}_g \subset \mathcal{RS}_g$.

(2.5) Corollary. $\mathcal{J}_g \subset \mathcal{S}_g \subset \mathcal{S}_g^{(\text{big})}$.

The point is, or course, the existence of Prym varieties. Both results follow from

(2.6) Schottky-Jung Identities. For $(C,\mu) \in \mathcal{RM}_g$ with Prym variety $P(C,\mu)$, we have an equality (in $\mathbb{P}(U_{g-1})/G_{g-1}$):

$$\alpha(P(C,\mu)) = B(J(C),\mu).$$

(This equality can be lifted to $\mathbb{P}(U_{g-1})$ by being careful to choose the right level-(2,4) structure on P corresponding to a given one on J. In this form (2.6) is known as the Schottky-Jung proportionality.)

This is just an analytic expression of the Splitting Theorem (2.2):

$$\pi'_*(\pi^{*\prime}(\theta)) = \theta + \theta_\mu \quad ,$$

where the LHS is interpreted via the Kummer-Prym Identification Theorem (2.3), and the RHS via the general Kummer Identification Theorem (0.14) (applied to the divisor $\theta_{\frac{1}{2}\mu}$ for some (any) choice of $\frac{1}{2}\mu$).

§ 2.3 Jacobians are a component of Schottky

The title result of this section was proved by van Geemen:

(2.7) Theorem [vG1] $\mathcal{J}_g$ is an irreducible component of $\mathcal{S}_g$.

In the sequel we will need a small improvement, with similar proof:

(2.8) Theorem [D2] $\mathcal{RJ}_g$ is an irreducible component of $\mathcal{RS}_g$, hence $\mathcal{J}_g$ is an irreducible component of $\mathcal{S}_g^{(big)}$.

Both proofs are based on degeneration to the boundary of moduli space, so let us begin with recalling the Satake-Baily-Borel compactification

$$\overline{\mathcal{A}}_g^s \supset \overline{\mathcal{A}}_{g-1}^s \supset \ldots \supset \overline{\mathcal{A}}_1^s \supset \mathcal{A}_0$$

where $\mathcal{A}_0$ is a point and

$$\overline{\mathcal{A}}_k^s \setminus \overline{\mathcal{A}}_{k-1}^s \approx \mathcal{A}_k.$$

Its boundary, ∂, is therefore irreducible, and is just a compactification of $\mathcal{A}_{g-1}$. The corresponding compactifications of level moduli spaces have reducible boundaries. We need a more precise description of the boundary of $\mathcal{RA}_g$.

Consider a corank -1 degeneration in $\mathcal{RA}_g$ with general fiber (X,μ), and let $\lambda \in X_2$ be the vanishing cycle (reduced mod. 2). In terms of the $\mathbb{Z}/2\mathbb{Z}$-valued intersection pairing (= Weil pairing) on X_2 we have 3 possibilities:

I. $\lambda = \mu$
II. $\lambda \neq \mu$, $(\lambda,\mu) = 0$
III. $(\lambda,\mu) \neq 0$.

These give (at least) 3 boundary components $\partial^I, \partial^{II}, \partial^{III}$ of $\overline{\mathcal{RA}}^s_g$.

(2.9) Lemma ([D2],[vG2] The boundary of $\overline{\mathcal{RA}}^s_g$ has exactly 3 irreducible components, described as above. They are isomorphic to the Satake-Baily-Borel compactification of $\mathcal{A}_{g-1}$, $\mathcal{RA}_{g-1}$, $\mathcal{A}_{g-1}$ respectively.

The idea for proving the theorems is then to analyze the boundary behavior of Schottky.

(2.10) Proposition. $\partial(\mathcal{RS}_g) = \partial^I \cup \partial^{III} \cup i_{II}(\mathcal{RS}_{g-1})$, where

$$i_{II} : \mathcal{RA}_{g-1} \subset \overline{\mathcal{RA}}^s_{g-1} \approx \partial^{II} \hookrightarrow \overline{\mathcal{RA}}^s_g$$

is the natural inclusion.

The reason ∂^I and ∂^{III} are in $\overline{\mathcal{RS}}^s_g$ is that they are the boundary of the locus of products

$$\overline{\mathcal{RA}}^s_1 \times \mathcal{A}_{g-1},$$

which is in $\mathcal{RS}_g$ since

$$\beta(X\times Y,\pi_1^*\mu) = \beta(X,\mu) \times \alpha(Y),$$

and the latter becomes $\alpha(Y)$ if $(X,\mu) \in \overline{\mathcal{RA}}_1^s$. ∂^{II} is not in this locus (since $\partial^{II}\overline{\mathcal{RA}}_1^s$ is empty!), and van Geemen shows that

$$\overline{\mathcal{RJ}}_g \cap \partial^{II} = i_{II}(\mathcal{RJ}_{g-1}).$$

The argument is now concluded by an induction. For Theorem (2.7), we need to show that the tangent cone to $\overline{\mathcal{J}}_g^s$ at a point $J(C)$ of $\mathcal{J}_{g-1} \subset \partial(\overline{\mathcal{J}}_g^s)$ is an irreducible component of the tangent cone to $\overline{\mathcal{J}}_g^s$ there. The latter is the intersection of the tangent cones to $\overline{\mathcal{RJ}}_g^s$ at the points $(J(C),\mu)$ for $\mu \in J_2 \setminus 0$, so it suffices to show the corresponding statement at a Jacobian point of any one of the 3 lifts $\partial^I,\partial^{II},\partial^{III}$. van Geemen does this at ∂^I. The picture is as follows:

- For $X \in \mathcal{A}_{g-1} \subset \partial\overline{\mathcal{A}}_g^s$, the projectivized tangent cone $\mathbb{P}T_X\overline{\mathcal{A}}_g^s$ is the Kummer variety $K(X) := X/(\pm 1)$.

- Let X denote also the corresponding point of $\partial^I\overline{\mathcal{RA}}_g^s$. Then $\mathbb{P}T_X\overline{\mathcal{RA}}_g^s$ maps isomorphically (by the forgetful map $\overline{\mathcal{RA}}_g^s \longrightarrow \overline{\mathcal{A}}_g^s$) to $\mathbb{P}T_X\overline{\mathcal{A}}_g^s \approx K(X)$.

- When $X = J(C)$, the subvariety

$$\mathbb{P}T_{J(C)}\overline{\mathcal{J}}_g^s \subset \mathbb{P}T_{J(C)}\overline{\mathcal{A}}_g^s$$

is a surface, the Abel-Jacobi image of $C - C$ in $K(X)$. (Ditto for $\mathbb{P}T_{J(C)}\overline{\mathcal{RJ}}_g^s \subset \mathbb{P}T_{J(C)}\overline{\mathcal{RA}}_g^s$.)

- For any $X \in \mathcal{A}_{g-1} \subset \partial^I\overline{\mathcal{A}}_g^s$, $\mathbb{P}T_X\overline{\mathcal{RJ}}_g^s$ can be computed by pulling back $T(\text{image } \alpha)$. It turns out to be the base locus, in $K(X)$, of the linear system

$$\Gamma_{00} := |\mathcal{O}_X(2\Theta)\otimes(\mathcal{I}_{\{0\}})^{\otimes 4}|$$

$$= \{s \in |\mathcal{O}_X(2\Theta)| \quad | \operatorname{mult}_0(s) \geq 4\}.$$

By Welters' Theorem (1.5), this base locus is known when $X = J(C)$ is a non-hyperelliptic Jacobian: it is again the surface $C - C$, except in genus 4 when it contains additionally the point $\pm(T_0 - T_1) \in K(X)$, where T_0, T_1 are the g^1_3's on X. In any case, $C - C$ is a component of the base locus, proving (2.7).

In proving (2.8) we do not have the freedom to switch boundary components, so we must work at ∂^{II}. The map $K(X) \longrightarrow \mathbb{P}(\Gamma^*_{00})$ given by the linear system Γ_{00} is then replaced by a projected Kummer map

$$K(\tilde{X}) \xrightarrow{\pi\circ\chi} \mathbb{P}(U_{g-2})$$

where $\tilde{X} \longrightarrow X$ is the double cover determined by μ, and

$$\pi : U_{g-1} \longrightarrow U_{g-2}$$

is the natural projection onto an eigenspace. The proof requires a second "blowup" (i.e. computation of tangent cone to the tangent cone), and is then reduced to an analogue of Welters' Theorem, a question on the linear system $|2\Theta_P|$ on a Prym.

§ 2.4. <u>Conjectures</u>

Unfortunately, $\mathcal{RS}_g$ does have components other than Jacobians. We have already noted that

$$\mathcal{RS}_g \supset \mathcal{RA}_1 \times \mathcal{A}_{g-1} \quad ,$$

and more generally

$$\mathcal{RS}_g \supset \mathcal{RJ}_{g-k} \times \mathcal{A}_k \quad , \qquad k \geq 4.$$

(For $k \leq 3$ the RHS is in the closure of $\mathcal{RJ}_g$.)

For many purposes, the toroidal compactifications $\bar{\mathcal{A}}_g^t$ [AMRT], and especially Voronoi's, are more convenient than the highly-singular $\bar{\mathcal{A}}_g^s$. In corank 1 (i.e. at generic points of the boundary components), a toroidal compactification looks like the blowup of $\bar{\mathcal{A}}_g^s$ along its boundary. We thus have

$$\partial\bar{\mathcal{A}}_g^t \sim \mathcal{X}_{g-1} \quad ,$$

and $\partial(\overline{\mathcal{RA}}_g^t)$ has 3 components with analogous descriptions.

In the toroidal version, the symmetry of ∂^I and ∂^{III} breaks: if we define

$$\overline{\mathcal{RS}}_g^t := \bar{\beta}^{-1} \quad (\text{image } \bar{\alpha})$$

for appropriate extensions $\bar{\alpha},\bar{\beta}$ of α,β, then $\overline{\mathcal{RS}}_g^t$ contains $\partial^I\overline{\mathcal{RA}}_g^t$ but not $\partial^{III}\overline{\mathcal{RA}}_g^t$. The point is that β extends to the Satake compactification near ∂^I, but only to the toroidal compactification near ∂^{III}. (This can be seen already on the Prym level. The Prym map $\mathcal{P} : \mathcal{RM}_g \longrightarrow \mathcal{A}_{g-1}$ extends to $\bar{\mathcal{P}} : \overline{\mathcal{RM}}_g \longrightarrow \bar{\mathcal{A}}_{g-1}$, where $\overline{\mathcal{RM}}_g$ is a stable-curve compactification. Its boundary components ∂^I, ∂^{III} map to $\mathcal{M}_{g-1}$ with 2-dimensional fibers. The extension of $\mathcal{P}$ to ∂^I depends only on the image point in $\mathcal{M}_{g-1}$ (Wirtinger's reducible double covers), but the extension to ∂^{III} depends on the fiber (Fay's double covers with 2 branch points) (cf. §4.2). This implies the corresponding statements for β, since by the Schottky-Jung Identities (2.6), $\beta = \alpha \circ \mathcal{P}$.) The upshot is that $\overline{\mathcal{RS}}_g^t$ contains $\partial^I\overline{\mathcal{RA}}_g^t$, but is only guaranteed to contain the zero-section of $\partial^{III}\overline{\mathcal{RA}}_g^t \sim \mathcal{X}_{g-1} \longrightarrow \mathcal{A}_{g-1}$ over a general ppav (and a surface in $K(X)$ for $X = J(C)$ a Jacobian, by the previous analysis of $\mathcal{P}$).

Finally, we will see in Chapter 5 that $\mathcal{RS}_5$ contains another component $\mathcal{RC}^0$, the moduli space of (intermediate Jacobians of) cubic threefolds with an even point of order 2. Assembling the pieces, we arrive at what we consider to be the natural formulation of the Schottky-Jung problem:

(2.11) <u>Conjecture.</u> The Schottky locus equals

$$\overline{\mathcal{RS}}_g = \overline{\mathcal{RJ}}_g \cup \partial^I\overline{\mathcal{RA}}_g \cup (\overline{\mathcal{RC}}^0 \times \overline{\mathcal{A}}_{g-5}) \cup \bigcup_{k\geq 4} \overline{\mathcal{RJ}}_{g-k} \times \overline{\mathcal{A}}_k \quad ,$$

where $\overline{}$ denotes (Voronoi's) toroidal compactification.

(2.12) <u>Corollary (of the conjecture).</u> $\mathcal{S}_g = \mathcal{J}_g$

(The conjectured components other than $\mathcal{RJ}_g$ do not contain a complete fiber of $\mathcal{RA}_g$ over $\mathcal{A}_g$.)

If another component of $\mathcal{RS}_g$ is discovered, the conjecture will of course need to be modified. As van Geemen pointed out though, the normal direction at $X \in \mathcal{J}_4$ to $\mathcal{A}_4 \sim \partial^I\overline{\mathcal{RA}}_5^s$ along the locus $\mathcal{RC}^0$ of cubic threefolds is given precisely by the difference of trigonal bundles $\pm(T_0 - T_1) \in K(X)$ [Co]. Since this is the only exception to Welters' Theorem, one hopes that $\mathcal{RC}^0$ is the only non-trivial component of $\mathcal{RS}_g$ other than Jacobians. (Of course, it is still possible that components exist which do not meet $\partial^I\mathcal{RJ}_g$, or meet it tangentially to one of the known components.)

In [vGvdG], van Geemen and van der Geer made (more or less) the following 4 conjectures (our versions of (2.14), (2.16) are slightly stronger):

<u>Conjectures [vGvdG]</u>

(2.13) The base locus of Γ_{00} in a Jacobian $J(C)$ is the surface $C - C$.

(2.14) The base locus of Γ_{00} in an indecomposable non-Jacobian X is (0).

(2.15) The intersection $\chi_X(X) \cap \overline{(\text{image } \alpha)}$ in a Jacobian $X = J(C)$ is the surface $\frac{1}{4}(C - C)$.

(2.16) The intersection $\chi_X(X) \cap \overline{(\text{image } \alpha)}$ in an indecomposable non-Jacobian X is (0).

Conjecture (2.13), with a slight modification, has since become Welters' Theorem. The base locus of Γ_{00} can also be described as the intersection

$$\chi_X(X) \cap T_X(\text{image } \alpha),$$

hence the analogy between the two pairs of conjectures.

(2.17) <u>Proposition.</u> The [vGvdG] conjectures follow from (2.11).

Indeed, these conjectures express the fact that, at Jacobian and non-Jacobian points of $\partial^I \overline{\mathcal{RA}}_g^t$ and $\partial^{III} \overline{\mathcal{RA}}_g^s$, the tangent cone to $\overline{\mathcal{RP}}_g$ is the tangent cone to the known components in the RHS of (2.11). (X must be assumed indecomposable to avoid the stupid components in (2.11).)

By considering the behavior of $\mathcal{RP}_g$ at ∂^{II}, we can make <u>one</u> more conjecture (recall that $\partial^{II}\mathcal{RP}_g$ is $\mathcal{RP}_{g-1}$, not all of $\mathcal{RA}_{g-1}$):

(2.18) <u>Conjecture.</u> Let $\tilde{C} \longrightarrow C$ be an unramified double cover with Prym P. We have maps

$$\text{Kummer} \quad \chi_P : P \longrightarrow \mathbb{P}(U_{g-1})/G_{g-1}$$

$$\text{Projected Kummer} \quad \pi \circ \chi_{\tilde{J}} : \tilde{J} \longrightarrow \mathbb{P}(U_{g-1})/G_{g-1}.$$

Then the intersection of the images is the image of $S^2\tilde{C}/i$, which maps to $K(\tilde{J})$, $K(P)$ by the Abel-Jacobi, Abel-Prym maps respectively.

If we believe Conjecture (2.11) as a scheme-theoretic statement, we get stronger versions of the conjectures. For instance, we "blow up" Conjectures (2.13), (2.14) at 0 : for any ppav X, let

$$\Gamma_{000} := \langle s \in \Gamma_{00} \mid \mathrm{mult}_0(s) \geq 6 \rangle.$$

Taking fourth-order terms gives an exact sequence

$$0 \longrightarrow \Gamma_{000} \longrightarrow \Gamma_{00} \longrightarrow |\mathcal{O}_{\mathbb{P}^{g-1}}(4)|,$$

so we can think of Γ_{00}/Γ_{000} as a linear system of quartics on $\mathbb{P}T_0X \approx \mathbb{P}^{g-1}$. From (2.11) we deduce:

(2.19) <u>Conjecture.</u> The base locus of the linear system Γ_{00}/Γ_{000} of quartics in $\mathbb{P}^{g-1} \approx \mathbb{P}T_0X$ is the canonical curve $\phi(C) \subset \mathbb{P}^{g-1}$, if $X = J(C)$ is a Jacobian, and is empty if X is an indecomposable ppav which is not a Jacobian.

The case of Jacobians follows from Welters' Theorem. For non-hyperelliptic curves it gives a very explicit prescription for recovering a curve from its Jacobian.

CHAPTER 3
Novikov

The theta function of a Jacobian satisfies a family of differential equations ("KP") which yield the best answer to the Schottky problem to date. The geometric explanation of these equations is based on the trisecants of a Jacobian Kummer variety; we discuss this in §3.2. Novikov's Conjecture (= Shiota's Theorem), saying that an abelian variety whose theta function satisfies KP is either a Jacobian or a product, is seen in §3.1 to follow from a more general conjecture which is in turn equivalent to Conjecture (2.19). We conclude with a brief description of the work of Beauville and Debarre which shows that the Novikov locus, of ppav's satisfying the KP equation (or various analogues), is contained in the Andreotti-Mayer locus, and in particular it contains the locus $\mathcal{J}_g$ of Jacobians as a component.

§ 3.1 More conjectures

Our starting point in this section is Conjecture (2.19), itself a corollary of Conjecture (2.11). We interpret it first in terms of linear differential relations satisfied by the vector-valued second-order theta function

$$\vartheta_2 : \mathbb{H}_g \times \mathbb{C}^g \longrightarrow U_g \approx \mathbb{C}^{2^g}$$

whose projectivization gives the Kummer map χ, and then in terms of non-linear differential equations satisfied by ϑ itself.

(3.1) Conjecture (Differential Characterization of Jacobians) An inedecomposable ppav X is a Jacobian if and only if its second-order theta function satisfies a constant-coefficient linear differential relation (i.e. polynomial in constant vector fields on X) of the form

$$((D_1)^4 + (\text{lower order terms}))\, \vartheta_2(\Omega,z)|_{z=0} = 0,$$

where Ω is any period matrix for X (i.e. $\Omega \in \mathbb{H}_g$ maps to $X \in \mathcal{A}_g$) and D_1 is a constant vector field on X.

This conjecture is simply a reformulation of (2.19). An element of $\Gamma = |\mathcal{O}_X(2\Theta)|$ is a linear combination of the entries of θ_2; it is in Γ_{00} if and only if all derivatives of order < 4 of this combination vanish at 0. Hence all the quartics in Γ_{00}/Γ_{000} vanish at some $D_1 \in T_0X$ if and only if $D_1^4\theta_2(\Omega,0)$ is a linear combination of lower order operators applied to $\theta_2(\Omega,0)$.

It is now natural to ask for the explicit form of the differential relations satisfied by Jacobian theta functions. Since the base locus of Γ_{00}/Γ_{000} in $J(C)$ is the canonical curve $\phi(C)$, we know that these equations are parametrized by C. We will find their explicit form in §3.2:

(3.2) <u>Proposition (Differential Relations for Jacobian Theta Functions)</u>. Let Ω be a period matrix of a Jacobian $X = J(C)$. Then $\vartheta_2(\Omega,0)$ satisfies precisely a one-dimensional family of inequivalent differential relations of the form (3.1). This family is parametrized by C; the equation corresponding to $p \in C$ is of the form

$$(D_1^4 - D_1D_3 + D_2^2 + d)\,\vartheta_2(\Omega,z)\big|_{z=0} = 0,$$

where d is a scalar constant, and the constant vector fields D_1, D_2, D_3 are determined by their values at AJ(p) (image of p under Abel-Jacobi), where they span the osculating line, plane and solid to AJ(C).

(3.3) <u>Corollary (Novikov's Conjecture, Dubrovin's Form)</u> An indecomposable ppav X is a Jacobian if and only if its second-order theta function satisfies a differential relation of the form (3.2).

This follows immediately from (3.1) and (3.2). Together the conjectures say that if ϑ_2 satisfies any equation of type (3.1) then we are on a Jacobian and the equation is of the form (3.2). Novikov's Conjecture has been proved by Shiota [Sh], but (3.1) is open.

The differential relations (3.1), (3.2) satisfied at 0 by the vector-valued ϑ_2 can be converted to a non-linear differential equation satisfied by the (scalar valued) theta function. This follows immediately from Riemann's Quadratic Identity (0.13):

$$\vartheta(z+w)\vartheta(z-w) = \sum_{\sigma\in(\mathbb{Z}/2\mathbb{Z})^g} \vartheta_2[\sigma](z)\ \vartheta_2[\sigma](w).$$

We treat one of the variables, say w, as a constant, and apply a differential operator to both sides, then evaluate at $z = 0$; this gives a differential expression in $\vartheta(w)$, on the left, and on the right a linear combination of the entries of the vector obtained by applying the operator to ϑ_2 at $z = 0$. For instance, (3.2) becomes:

$$\begin{aligned}(3.4)\qquad & D_1^4\vartheta\cdot\vartheta - 4D_1^3\vartheta\cdot D_1\vartheta + 3(D_1^2\vartheta)^2 - D_1D_3\vartheta\cdot\vartheta + D_1\vartheta\cdot D_3\vartheta \\ & + D_2^2\vartheta\cdot\vartheta - (D_2\vartheta)^2 + \tfrac{1}{2}d\vartheta^2 = 0.\end{aligned}$$

This is known as Hirota's bilinear form of the KP (= Kadomtsev-Petviashvilli) equation. The standard form of this differential equation is:

$$(3.5)\qquad (u_{xxx} + uu_x - u_t)_x + u_{yy} = 0.$$

Direct substitution shows that (3.4) for ϑ is equivalent to the KP equation (3.5) for $u := (\log\vartheta)_{xx}$.

§ 3.2. Trisecants and the KP hierarchy

The Kummer variety of a Jacobian, as embedded in $\mathbb{P}(U_g)$, has a four-dimensional family of trisecant lines. The KP equation (3.2), as well as a whole hierarchy of equations satisfied by Jacobian theta functions, express limiting cases of the existence of these trisecants. Our presentation here is based on ideas of Gunning, Welters and Arbarello-De Concini.

(3.6) Lemma. Let a,b,c,d be points of a curve C. The various translates of the divisor $\Theta \in J(C)$ satisfy the following inclusions:

(1) $\Theta \cap \Theta_{a-b} \subset \Theta_{a-c} \cup \Theta_{d-b}$

(2) $(\theta_{a+b-c-d} \cup \theta) \supset (\theta_{a-d} \cup \theta_{b-c}) \cap (\theta_{a-c} \cup \theta_{b-d})$

(3)

$$\left[\theta_{\frac{a+b-c-d}{2}} \cup \theta_{\frac{-a-b+c+d}{2}}\right] \supset \left[\theta_{\frac{a-b+c-d}{2}} \cup \theta_{\frac{-a+b-c+d}{2}}\right] \cap \left[\theta_{\frac{a-b-c+d}{2}} \cup \theta_{\frac{-a+b+c-d}{2}}\right],$$

where the choices of halves are compatible, i.e. we fix one of the 2^{2g} values of $\frac{a+b-c-d}{2}$, and determine all other expressions accordingly:

$$\frac{a-b+c-d}{2} := \frac{a+b-c-d}{2} - b + c, \quad \frac{a-b-c+d}{2} := \frac{a+b-c-d}{2} - b + d, \text{ etc.}$$

Proof.

(1) follows from Riemann-Roch. (2) follows from (1) by expanding the RHS as union of four intersections: the inclusion of each in the LHS is equivalent to (1), with the letters permuted, after translation. (3) is equivalent to (2) via translation by the fixed value of $\frac{a+b-c-d}{2}$.

QED

This lemma is classical (Mumford [M3] attributes it to Weil), but its interpretation via trisecants was first noticed by Fay:

(3.7) <u>Corollary [F]</u> For $a,b,c,d \in C$, the three points

$$\chi\left(\frac{a+b-c-d}{2}\right), \ \chi\left(\frac{a-b+c-d}{2}\right), \ \chi\left(\frac{a-b-c+d}{2}\right)$$

of the Kummer are collinear. (The halves must be compatible as in (3.6)(3).)

The corollary is just a restatement of (3.6)(3), using the Kummer Identification Theorem (0.14).

We see that a Jacobian Kummer has a 4-dimensional family of trisecants. The group $J_2(C)$ of points of order 2 acts linearly on $\mathbb{P}(U_g)$ inducing translation on $K(J_2(C))$, hence acts on the variety of trisecants. Let S be the quotient. It is clear from (3.7) that S is birationally equivalent to S^4C, and an easy additional computation shows that

$$S \approx S^4 C$$

biregularly. Let us see what happens to a trisecant as we bring the points a,b,c,d together:

Choose a point $a \in C$, and write down the Taylor expansion in $\mathbb{C}^g$ of the Abel-Jacobi map near a, in terms of a coordinate t on C near a:

$$\text{(3.8)} \qquad AJ(t) = AJ(a) + tD_1 + t^2D_2 + t^3D_3 + \ldots$$

where $D_1, D_2, \ldots$ are constant vectors in $\mathbb{C}^g$. (We can also think of them as translation-invariant vector fields on J(C).)

For general a,b,c,d, (3.7) says that the 3 vectors in U_g:

$$\vartheta_2\left(\frac{a+b-c-d}{2}\right),\ \vartheta_2\left(\frac{a-b+c-d}{2}\right),\ \vartheta_2\left(\frac{a-b-c+d}{2}\right)$$

are linearly dependent. Let us bring two of the points together, say $c \longrightarrow a$. The 3 vectors become:

$$\vartheta_2\left(\frac{b-d}{2}\right),\ D_1\vartheta_2\left(\frac{b-d}{2}\right),\ \vartheta_2\left(\frac{b+d}{2} - a\right).$$

Next we may proceed in two different ways:

(A) Bring $d \longrightarrow b$. The vectors become

$$\vartheta_2(0),\ D_1D_1'\vartheta_2(0),\ \vartheta_2(a-b)$$

where D_1' is the first term in the Taylor expansion of AJ(C) near b. (Recall that ϑ_2 is even, so its first derivatives at 0 vanish.)

Now let us take b near a, corresponding to the value t of the coordinate at a. Differentiating (3.8) gives.

$$D_1' = D_1 + 2tD_2 + 3t^2D_3 + \ldots$$

while $\vartheta_2(a-b)$ becomes

$$\vartheta_2(0)+t^2D_1^2\vartheta_2(0)+2t^3D_1D_2\vartheta_2(0)+t^4(D_2^2+2D_1D_3+D_1^4)\vartheta_2(0)+\ldots$$

or, subtracting $\vartheta_2(0) + t^2D_1D_1'\vartheta_2(0)$:

$$t^4[(D_2^2 - D_1D_3 + D_1^4)\vartheta_2+\ldots].$$

Setting this to equal a linear combination of $\vartheta_2(0)$ and $D_1D_1'\vartheta_2(0)$ gives an infinite sequence of differential relations obtained by equating successive powers of t to 0. The leading term (coefficient of t^4) gives exactly Proposition (3.2). (D_3 may have to be replaced by a linear combination of D_3 and D_1.)

(B) In the previous computation we brought d to b, i.e. considered fourtuples of the form (a,b,a,b), resulting in the trisecant becoming a tangent line at 0 meeting K(J(C)) elsewhere (at $\vartheta_2(a-b)$). Instead, we may bring d to a, i.e. consider fourtuples (a,b,a,a). The limiting trisecants now become flexes of the Kummer, at the point $\vartheta_2(\frac{a-b}{2})$, i.e. we obtain the linear dependence of

$$\vartheta_2(\tfrac{a-b}{2}),\ D_1\vartheta_2(\tfrac{a-b}{2}),\ (D_2+\tfrac{1}{2}D_1^2)\vartheta_2(\tfrac{a-b}{2}).$$

Again, we obtain an infinite set of differential relations satisfied by Θ_2. The first of these is again (3.2), but the relation of this sequence to the one described above is not clear. Arbarello and De Concini show in [AdC1] that this sequence of equations is a consequence of the "KP hierarchy", an infinite systems of PDE's, starting with (3.5), which can be interpreted as an infinite-dimensional completely integrable Hamiltonian system.

The fact that Jacobian theta functions satisfy the KP equation, indeed the KP hierarchy, was discovered by Krichever. That led Novikov to conjecture (3.3). Dubrovin observed in [Du] that the Hirota form (3.4) of the equations is equivalent to the differential relation (3.2), and proved that Jacobians form a component of the locus of ppav's whose theta functions satisfy (3.2). Mulase [Mu] showed that an indecomposable ppav whose theta function satisfies the KP hierarchy is a Jacobian. Arbarello and De Concini showed [AdC1], based on earlier work of Gunning and Welters [We2], that a finite subset of this hierarchy suffices (the equations in (B) above corresponding to powers of t up to $6^g \cdot g! + 1$). The Novikov Conjecture itself was proved by Shiota [Sh]; a simplified proof is in [AdC2]. Various analogues have been proposed, but remain open. For instance, Welters asks in [We2] whether the existence of one trisecant of the Kummer variety forces it to come from a curve.

§ 3.3. Andreotti-Mayer vs. Novikov

(3.9) Theorem [BD] $A \in \mathcal{A}_g$ is in the Andreotti-Mayer locus $\mathcal{AM}_g$ if it satisfies any of the following conditions:

(1) There are distinct points $x,y,z \in A$ such that

$$\Theta \cap \Theta_z \subset \Theta_x \cup \Theta_y.$$

(2) The Kummer variety $K(A)$ has a trisecant.

(3) The theta function ϑ_A satisfies the KP equation (3.2).

Consider the map

$$R : A \setminus (0) \longrightarrow \mathrm{Div}(\Theta)$$

sending $a \in A \setminus (0)$ to the divisor $\Theta \cap \Theta_a$ in Θ. This extends to a morphism

$$R : \tilde{A} \longrightarrow \mathrm{Div}(\Theta)$$

where $\tilde{A}$ is the blowup of A at 0: a point in the exceptional divisor, corresponding to a vector field D_1 on A, goes to the divisor

$$\{z \in A | \vartheta(z) = D_1\vartheta(z) = 0\} \subset \Theta.$$

The theorem of Beauville and Debarre follows from:

(3.9 bis) Theorem [BD]. Assume A satisfies:

(0) The divisor $R(a) \subset \Theta$ is reducible for some $a \in \tilde{A}$.

Then either $A \in \mathcal{AM}_g$ or A contains an elliptic curve E such that $E \circ \Theta = 2$.

Condition (1) implies (0) for $a = z \in A \setminus (0)$. By (0.14), condition (2) is equivalent either to (1) or to a limiting form, so it also implies (0). Fianally, (3) implies (0) for a in the exceptional divisor, corresponding to the vector D_1 in (3.2). This is immediate from Hirota's version (3.4) of (3.2): setting $\vartheta = D_1\vartheta = 0$ we get a product,

$$0 = 3(D_1^2\vartheta)^2 - (D_2\vartheta)^2 = (\sqrt{3}D_1^2\vartheta - D_2\vartheta)(\sqrt{3}D_1^2\vartheta + D_2\vartheta).$$

The idea for proving (3.9 bis) is that if $A \notin \mathcal{AM}_g$ then Θ is singular in codimension > 3, hence is locally factorial; the reducbile $R(a)$ is thus the sum $C + C'$ of two effective Cartier divisors in Θ. These in turn come from divisors on A, and the resulting configuration forces the existence of E. Finally, the existence of E can be ruled out assuming conditions (1), (2) or (3).

CHAPTER 4

Andreotti-Mayer in low genus

In this chapter we present the results of [M2], [B] and [D1]: $\mathcal{AM}_4$, the first non-trivial Andreotti-Mayer locus, consists of $\mathcal{J}_4$ and another divisor (ϑ_{null}) in $\mathcal{A}_4$; $\mathcal{AM}_5$ consists of $\mathcal{J}_5$, products, and Pryms of bielliptic curves. The idea is to study $\mathrm{Sing}(\theta)$ for Prym varieties and their degenerations, and to use the dominance of the Prym map to $\mathcal{A}_g$ for $g \leq 5$.

§ 4.1 Sing(θ) for a Prym

(4.1) Theorem [M2] Let $P = P(C,\mu)$ be the Prym variety of $(C,\mu) \in \mathcal{RM}_g$. If P is in the Andreotti-Mayer locus $\mathcal{AM}_{g-1}$, then (C,μ) is one of the following:

(a) hyperelliptic

(b) trigonal

(c) bielliptic (i.e. branched double cover of an elliptic curve)

(d) $g = 5$, C has a vanishing even thetanull L, and $L \otimes \mu$ is even (i.e. L satisfies $L^{\otimes 2} = \omega_C$, $h^0(L) = 2, h^0(L \otimes \mu) = 0$).

(e) $g = 6$, C is a plane quintic curve, and μ (i.e. $h^0(\mu \otimes \mathcal{O}_C(1))$) is even.

The converse is also true. In case (a), P is a hyperelliptic Jacobian, in $\mathcal{N}^{g-4}_{g-1}$, or a product of two, in $\mathcal{N}^{g-3}_{g-1}$. In cases (b) and (e), P is a Jacobian (cf. Corollary (4.12) for (b)). In cases (c), (d), P is not a Jacobian, but it follows from the description of $\mathrm{Sing}(\Xi)$ below that P is still in $\mathcal{AM}_{g-1}$.

The starting point for the proof is Wirtinger's Theorem (2.1):

$$\tilde{\theta} \cap P = 2\Xi.$$

Let $\tilde{C} \longrightarrow C$ be the double cover given by μ. After translation, we can think of a point of P as given by a line bundle L on $\tilde{C}$

satisfying $Nm(L) = \omega_C$ (and a parity condition). There are two ways that L can represent a singular point of Ξ:

Type (1): $\mathrm{mult}_L(\tilde{\theta}) \geq 4$.
Type (2): $\mathrm{mult}_L(\tilde{\theta}) = 2$, and $T_L P \subset T_L \tilde{\theta}$.

(4.2) Lemma In any component of $\mathrm{Sing}(\Xi)$ whose dimension is $\geq g - 5$, the generic point is of type (2).

This lemma allows Mumford to ignore type (1) singularities. Type (2) singularities can be described directly, and the question is transformed to finding all curves C on which

$$\dim (W^1_d) > d - 4$$

for some $d \leq g - 2$. By a theorem of Martens and Mumford ([M2, appendix] or [ACGH, Ch. IV, Theorems (5.1), (5.2)]), this occurs only for the exceptional curves listed in the theorem.

§ 4.2. Prym is proper

The Prym map

$$\mathcal{P} : \mathcal{RM}_g \longrightarrow \mathcal{A}_{g-1}$$

is not proper, but can be made proper as follows. Let $\overline{\mathcal{RM}}_g$ denote the stable-curve compactification of $\mathcal{RM}_g$. By the universal extension property of the Satake-Baily-Borel compactification $\overline{\mathcal{A}}^s_{g-1}$ [Bo], there is an extension

$$\overline{\mathcal{P}} : \overline{\mathcal{RM}}_g \longrightarrow \overline{\mathcal{A}}^s_{g-1}.$$

We then define $(\mathcal{RM}_g)_{\text{allowable}}$ to be the inverse image in $\overline{\mathcal{RM}}_g$ of the open subset $\mathcal{A}_{g-1}$. The resulting map

$$\mathcal{P}_{\text{allowable}} : (\mathcal{RM}_g)_{\text{allowable}} \longrightarrow \mathcal{A}_{g-1}$$

is then a proper extension of $\mathcal{P}$.

It is more interesting to interpret this extension geometrically, i.e. to describe which degenerate double covers are allowable. There are 5 "types" of boundary components: first we have the 3 components ∂^{I}, ∂^{II}, ∂^{III} of $\partial\mathcal{R}\mathcal{M}_g$ which are the restrictions of the corresponding components of $\partial\mathcal{R}\mathcal{A}_g$. Additionally, we have two families of boundary components consisting entirely of covers of reducible base-curves:

• ∂_k, for $1 \leq k \leq g-1$, parametrizes double covers $\tilde{C} \longrightarrow C$ where C is reducible:

$$C = X \cup_p Y$$

with Y,X of genera $k,g-k$ respectively, meeting transversally at p, and $\tilde{C}$ is the double cover corresponding to a point of order 2 $\mu \in J_2(Y) \setminus (0)$.

• $\partial_{k,g-k}$, for $1 \leq k \leq g-k$, parametrizes reducible covers with C as above but μ supported on both X and Y.

It is quite easy to see that ∂^{I}, ∂^{III} and ∂_k are allowable, while ∂^{II}, $\partial_{k,g-k}$ are not. The degenerate double covers in ∂^{I} ("Wirtinger covers") are of the form

$$C := X/(p \sim q), \quad \tilde{C} := (X_0 \coprod X_1)/(p_o \sim q_1, q_0 \sim p_1)$$

where $X \in \mathcal{M}_{g-1}$, $p,q \in X$, and X_0, X_1 are two copies of X. The limiting Prym in this case is just $J(X)$. The degenerate double covers in ∂^{III} ("Beauville covers") are of the form

$$C := X/(p \sim q), \quad \tilde{C} := \tilde{X}/(\tilde{p} \sim \tilde{q})$$

where $X \in \mathcal{M}_{g-1}$ and $\tilde{X} \longrightarrow X$ is ramified at p,q. The limiting Prym $P(\tilde{C}/C)$ is just $P(\tilde{X}/X)$. (Fay showed in [F] that the Pryms of double covers with two branch points are ppav's.) A ∂_k-cover is of the form

$$C := Y \cup_p X, \quad \tilde{C} := X_0 \cup_{p_0} \tilde{Y} \cup_{p_1} X_1$$

and its limiting Prym is $J(X) \times P(\tilde{Y}/Y)$.

This takes care of "corank 1 degenerations" of double covers, but the same ideas extend to arbitrary degenerations: any stratum of $\overline{\mathcal{RM}}_g$ is locally the intersection of several boundary components (some of these components have self-intersection), and the result is allowable iff only allowable components are involved. More explicitly:

(4.3) <u>Definition</u> A branched double cover $\tilde{C} \longrightarrow C$ of stable curves is:

(1) A stable Wirtinger degeneration if it is of the form

$$\begin{array}{c} \tilde{C} := (X_0 \amalg X_1)/(p_0 \sim q_1,\ q_0 \sim p_1) \\ \downarrow \\ C := X\,/\,(p \sim q), \end{array}$$

with X stable.

(2) An allowable reducible degeneration, if $C = Y \cup_{\underline{p}} X$ (where Y is stable, $X = \amalg_{i\in I} X^i$ is the disjoint union of stable curves X^i, and the glueing set $\underline{p} = \{p^i\}_{i\in I}$ contains one point in each X^i) and the corresponding cover is $\tilde{C} = X_0 \amalg_{\underline{p}_0} \tilde{Y} \amalg_{\underline{p}_1} X_1$ with $X_0 \approx X_1 \approx X$ and $\tilde{Y} \longrightarrow Y$ any stable double cover.

(3) A stable Beauville degeneration if the branch points of the map $\tilde{X} \longrightarrow X$ of normalizations are precisely the inverse images of the nodes.

The result is then:

(4.4) Theorem [B] A stable, branched double cover $\pi : \tilde{C} \longrightarrow C$ is allowable if and only if it is either

(a) a stable Wirtinger,

$$C = X/(p \sim q)$$

with X treelike (i.e. the graph of components of X is a tree); or:

(b) allowable reducible,

$$C = Y \cup_{\underline{p}} X, \qquad \tilde{C} = X_0 \cup_{\underline{p}_0} \tilde{Y} \cup_{\underline{p}_1} X_1$$

where each connected component of X is treelike, and where

$$\tilde{Y} \longrightarrow Y$$

is stable Beauville.

§ 4.3 Sing(θ) for generalized Pryms

Beauville has extended Mumford's analysis to allowable covers of singular curves:

(4.5) Theorem [B] Consider a stable Beauville degeneration $\tilde{C} \longrightarrow C$ with Prym $P = P(\tilde{C}/C) \in \mathcal{A}_{g-1}$.

(1) If $P \in \mathcal{N}^{g-3}_{g-1}$, C is either hyperelliptic or a union $C = C_0 \cup C_1$ with $\#(C_0 \cap C_1) = 2$.

(2) If $P \in \mathcal{N}^{g-4}_{g-1}$, C is either hyperelliptic or hyperelliptic with 2 points identified.

(3) If $P \in \mathcal{AM}_{g-1} = \mathcal{N}^{g-5}_{g-1}$, C is one of the following:

(a) trigonal

(b) hyperelliptic with two points identified

(c) bielliptic, $g \geq 6$

(d) $g = 5$, C has a vanishing theta null, $\tilde{C} \longrightarrow C$ is even

(e) $g = 6$, a plane quintic with even double cover.

(f) hyperelliptic with two pairs of points identified.

(g) $g = 5$, a genus-4 curve with vanishing theta null and with a pair of points identified.

(h),(i),(j), $C = C_0 \cup X_1$, $\#(X_0 \cap X_1) = 4$, and either:

(h) neither X_0 nor X_1 is rational.

(i) X_0 is rational, X_1 hyperelliptic of genus ≥ 3.

(j) X_0 is rational, X_1 is of genus 3, $\omega_{X_1} \approx \mathcal{O}_{X_1}(X_0 \cap X_1)$ (hence $g = 6$).

It was known already to Wirtinger [W] that $\mathcal{P} : \mathcal{RM}_g \longrightarrow \mathcal{A}_{g-1}$ is dominant for $g \leq 6$. (This is easiest to see by computing the differential of $\mathcal{P}$ along the locus ∂^I of "Wirtinger covers".) Combining with §4.2, one gets

(4.6) Lemma $\mathcal{P}_{allowable} : (\mathcal{RM}_g)_{allowable} \longrightarrow \mathcal{A}_{g-1}$ is surjective for $g \leq 6$.

One can therefore completely analyze $\mathcal{AM}_g$ for $g \leq 5$: By Theorem (4.4), anything in $\mathcal{AM}_g$ is either a Wirtinger Prym (which is a Jacobian or product of Jacobians), or a product, or a stable Beauville Prym which is therefore in the list (4.5). Going through the list, Beauville deduces:

(4.7) Corollary. $\mathcal{AM}_4$ has 2 irreducible components: $\mathcal{J}_4$ and the divisor θ_{null} of ppav's with a vanishing theta null.

(4.8) Corollary. All components of $\mathcal{AM}_5$ other than $\mathcal{J}_5$ are contained in the divisor θ_{null}.

(Unfortunately, ϑ_{null} is contained in $\mathcal{N}_5^0$ but not in $\mathcal{N}_5^1 = \mathcal{A}\mathcal{M}_5$.)

§ 4.4 The tetragonal construction

We describe a simple procedure, the tetragonal construction, which takes a tower

(4.9) $$\tilde{C} \xrightarrow{\pi} C \xrightarrow{f} \mathbb{P}^1,$$

where f is a 4-sheeted branched cover (i.e. C is tetragonal) and π is an unramified double cover, and yields two new towers

$(4.9)_i$ $$\tilde{C}_i \xrightarrow{\pi_i} C_i \xrightarrow{f_i} \mathbb{P}^1 \qquad i = 0,1$$

of the same type. Such a tower is uniquely determined by a representation ρ of $\pi_1(\mathbb{P}^1 \setminus \{\text{branch points}\})$ in the Weyl group WD_4 of the Dynkin diagram:

D_4 : 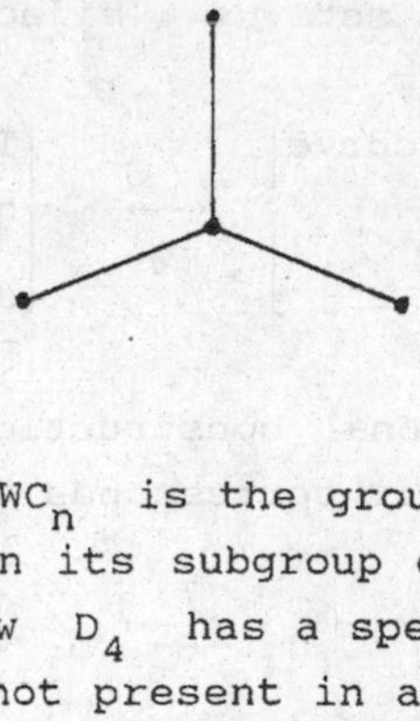

(In general, the Weyl group WC_n is the group of signed permutations of n letters, and WD_n in its subgroup of index 2 consisting of even signed permutations.) Now D_4 has a special automorphism α, of order 3, (120° rotation), not present in any other D_n. This gives an outer automorphism α of WD_4. Therefore representations of any gorup in WD_4 come in packages of three: ρ, $\alpha \circ \rho$, $\alpha^2 \circ \rho$. In particular, we get $(4.9)_i$ $(i = 0,1)$ starting with (4.9).

(More explicitly, starting with (4.9) we construct a

$(16 = 2^4)$-sheeted branched cover $f_*\tilde{C} \to \mathbb{P}^1$ with a natural involution. This breaks into two components, each of degree 8 over $\mathbb{P}^1$ and invariant under the involution, yielding $(4.9)_i$, cf. [D1].)

(4.10) Theorem [D1] The tetragonal construction commutes with the Prym map:

$$P(\tilde{C}/C) \approx P(\tilde{C}_0/C_0) \approx P(\tilde{C}_1/C_1).$$

Consider the special case where π in (4.9) is the split double cover. The 16-sheeted branched cover $f_*\tilde{C}$ then splits into 5 components of degrees 1,4,6,4,1 respectively over $\mathbb{P}^1$. The components of degree 4 make up $\tilde{C}_1 \to C_1$, which is isomorphic to $\tilde{C} \to C$. The remaning components give

$$\mathbb{P}^1 \coprod \tilde{T} \coprod \mathbb{P}^1 \longrightarrow T \coprod \mathbb{P}^1$$

where T is a trigonal curve and $\tilde{T}$ its double cover. One sees easily that this special case sets up a bijection

$$(4.11) \quad \left\{\begin{array}{l}C\text{, a tetragonal curve}\\ \\ \text{of genus } g\end{array}\right\} \longleftrightarrow \left\{\begin{array}{l}T\text{ ,a trigonal curve of genus}\\ g+1\text{ with an unramified}\\ \text{double cover } \tilde{T}\end{array}\right\}$$

This bijection, the trigonal construction, was described by Recillas [R]. Group theoretically it corresponds to the exceptional isomorphism

$$S_4 \xrightarrow{\sim} WD_3$$

which arises from the coincidence of Dynkin diagrams A_3, D_3. (In general, the symmetric group S_n is the Weyl group WA_{n-1}.) Theorem (4.10) thus yields:

(4.12) Corollary [R] If $(\tilde{T},T)$ corresponds to C via the trigonal construction, then $P(\tilde{T}/T) \approx J(C)$.

(In particular, this shows that case (b) in Mumford's Theorem (4.1) leads to Jacobians.)

The tetragonal construction (4.9) (though not Theorem (4.10)) can be deduced from the trigonal construction (4.11): starting with a trigonal T of genus $g+1$, choose a rank-2 isotropic subgroup of $(J(T))_2$, i.e. 3 points of order 2 μ, μ_0, μ_1 satisfying

(4.13) $$\mu + \mu_0 + \mu_1 = 0, \qquad (\mu, \mu_0) = 0 \in \mathbb{Z}/2\mathbb{Z}.$$

These points of order 2 determine double covers $\tilde{T}, \tilde{T}_0, \tilde{T}_1$ of T. Applying (4.11) we get 3 curves of genus g: C, C_0, C_1, each with a g^1_4. Finally, each of these curves comes with a point of order 2 in its Jacobian, hence a double cover: for C, this point is the common image in $J(C) \approx P(\tilde{T}/T)$ of μ_0 and μ_1.

Using the tetragonal construction, we can obtain many identifications among Prym varieties of special curves. As an illustration, let us consider double covers of bielliptic curves. If C^0, C^1 are branched double covers of an elliptic curve E with disjoint branch loci, we form the fiber product

$$\tilde{C} := C^0 \times_E C^1.$$

$\tilde{C}$ has 3 involutions $:\tau^i$, with quotient C^i, $(i = 0,1)$, and their composition $\tau := \tau^0 \circ \tau^1$ which is fixed-point free, yielding an unramified double cover $\tilde{C} \to C$ of a quotient curve C which is itself bielliptic. We say that $\tilde{C} \to C$ is a Cartesian cover.

Let $\mathcal{B} = \mathcal{B}_g$ be the moduli space of bielliptic curves of genus g. The space $\mathcal{R}\mathcal{B}_g$ of bielliptic curves with a double cover has $[\frac{g+1}{2}] + 1$ components:

- $\mathcal{RB}_i$, $1 \le i \le [\frac{g+1}{2}]$, consists of Cartesian covers $\tilde{C} \longrightarrow C$ arising from a pair of covers $C^0 \longrightarrow E$, $C^1 \longrightarrow E$ where C^0 is of genus i, C^1 of genus $g + 1 - i$.

- $\mathcal{RB}_0$ consists of all non-Cartesian covers.

Each of these components is $(2g - 2)$-dimensional. Since each bielliptic curve has a 1-dimensional family of g^1_4's, we have ample room to play with the tetragonal construction. The result:

(4.14) Proposition

(1) $\mathcal{P}(\mathcal{RB}_1)$ is $(2g - 2)$-dimensional and is also the locus of allowable Pryms of hyperelliptic curves with two pairs of points identified.

(2) $\mathcal{P}(\mathcal{RB}_i)$, for $i > 1$, is $(2g - 3)$-dimensional, and is also the locus of Pryms of reducible allowable covers $\tilde{C} \longrightarrow C$ where $C = X_0 \cup X_1$, $\tilde{C} = \tilde{X}_0 \cup \tilde{X}_1$, X is hyperelliptic of genus $i - 2$, X_1 is hyperelliptic of genus $g - 1 - i$.

(3) $\mathcal{P}(\mathcal{RB}_0) = \mathcal{P}(\mathcal{RB}_1)$.

In particular, when $g = 6$, Beauville's list becomes quite short. (The anouncement in [D1] is wrong. I thank O. Debarre for pointing this out.)

(4.15) Theorem. $\mathcal{AM}_5$ consists of the 12-dimensional locus $\mathcal{J}_5$ of Jacobians, the 11-dimensional locus $\mathcal{A}_1 \times \mathcal{A}_4$ of products, and the 3 loci $\mathcal{P}(\mathcal{RB}_i)$, $i = 1,2,3$, of Cartesian bielliptic Pryms (these have dimensions 10,9,9 repectively.)

CHAPTER 5
Schottky-Jung in low genus

In §5.1 we try to explain why the unexpected component $\mathcal{RC}^0$ (of intermediate Jacobians of cubic threefolds with even point of order 2) pops into the Schottky locus $\mathcal{RS}_g$ in genus 5 (hence also for $g \geq 5$). This explains our formulation of the Schottky-Jung Conjecture (2.11).

The case $g = 4$ of the conjecture amounts to Igusa's Theorem. In the last section we sketch a new proof of this result, based on the various symmeties of Pryms and thetas, and in the same spirit we outline our recent proof (yet unpublished) of the conjecture in genus 5.

§ 5.1. Symmetry of the theta maps

In this section we discuss an extension of the tetragonal construction to arbitrary curves. Let $Q \in \mathcal{M}_{g+1}$ be a curve, and

$$\langle 0, \mu_0, \mu_1, \mu_2 = \mu_0 + \mu_1 \rangle$$

an isotropic rank-2 subgroup of $J_2(Q)$. For $i = 0,1,2$ we have a Prym variety

$$P_i = P(Q,\mu_i) \in \mathcal{A}_g$$

and on it a uniquely determined semiperiod ν_i, image of any μ_j $(j \neq i)$ in P_i. The result is:

(5.1) Theorem [D3]. $\beta(P_i,\nu_i)$ is independent of $i = 0,1,2$.

This has many geometric applications. Taking $Q = T$ to be trigonal, we find ourselves in the situation of (4.14): each P_i is a Jacobian of a tetragonal curve, the three are related via the tetragonal construction, and this special case of (5.1) follows from (and is slightly weaker than) Theorem (4.10). Taking (Q,μ_0) to be a

Wirtinger (∂^I) degeneration, (5.1) becomes the Schottky-Jung identity (2.6).

Of interest to us is the case that Q is a plane quintic curve, μ_2 is an odd point of order 2, μ_0 (hence also μ_1) is even. By Theorem (4.1)(e), P_0 and P_1 are Jacobians of curves of genus 5, but P_2 is not.

Let $\mathcal{C} \subset \mathcal{A}_5$ be the 10-dimensional locus of Pryms of quintics with odd covers. (It is known ([CG], [M2], [T]) that this is precisely the locus of intermediate Jacobians of cubic threefolds.) Its lift to $\mathcal{R}\mathcal{A}_5$ splits into two components $\mathcal{R}\mathcal{C}^0$, $\mathcal{R}\mathcal{C}^1$, where $\mathcal{R}\mathcal{C}^0$ parametrizes precisely the pairs (P_2, ν_2) arising as above.

(5.2) <u>Theorem [D3].</u> The locus $\mathcal{R}\mathcal{C}^0$ of intermediate Jacobians of cubic threefolds with an even point of order 2 is a component of the Schottky locus $\mathcal{R}\mathcal{S}_5$.

The inclusion $\mathcal{R}\mathcal{C}^0 \subset \mathcal{R}\mathcal{S}_5$ follows immediately from the symmetry result (5.1) and the Schottky-Jung identities (2.6). The proof that $\mathcal{R}\mathcal{C}^0$ is actually a component was suggested by van Geemen. It is analogous to the proof that $\mathcal{R}\mathcal{J}$ is a component of $\mathcal{R}\mathcal{S}$, Theorems (2.7) and (2.8): the closure of $\mathcal{R}\mathcal{C}^0$ meets $\partial^I\mathcal{R}\mathcal{A}_5 \approx \mathcal{A}_4$ in the locus of Jacobians $\mathcal{J}_4$, and by a result of Collino [Co] the projectivized normal cone to ∂^I along $\mathcal{R}\mathcal{C}^0$ at $J(C) \in \mathcal{J}_4$ is given by the point

$$\pm(T_0 - T_1) \in K(J(C)),$$

where T_0, T_1 are the g^1_3's on C. This is precisely the exceptional case in Welters' Theorem (1.5): for generic C, the point $\pm(T_0 - T_1)$ is not in $C - C$, hence forms a component of the base locus of Γ_{00}, which proves the theorem.

The theorem implies, of course, that $\mathcal{C} \subset \mathcal{S}_5^{(big)}$. We will see below that $\mathcal{C}$ is not in $\mathcal{S}_5$.

§ 5.2. Schottky in genus ≤ 5

The theta map (0.12):

$$\beta : \mathcal{RA}_g \longrightarrow \mathbb{P}(U_{g-1})/G_{g-1}$$

extends to a proper map on an appropriate toroidal compactification,

$$\beta : \overline{\mathcal{RA}}_g^t \longrightarrow \mathbb{P}(U_{g-1})/G_{g-1}.$$

For $g \leq 5$ this map is surjective. Our strategy is to study the geometry of this map and to use it to completely describe $\overline{\mathcal{RS}}_g$.

(5.3) Theorem For $C \in \mathcal{M}_3$, the inverse image

$$\beta^{-1}(\alpha(J(C)))$$

consists of two copies of the Kummer $K(J(C))$: one is

$$\mathcal{P}^{-1}(J(C)) \subset \overline{\mathcal{RM}}_4,$$

the other is the fiber over $J(C) \in \mathcal{M}_3$ of the natural map

$$\begin{array}{ccc} \partial^I \overline{\mathcal{RA}}_4^t & \longrightarrow & \partial^I \overline{\mathcal{RA}}_4^s \\ \wr\wr & & \wr\wr \\ \overline{\mathcal{X}}_3 & \longrightarrow & \overline{\mathcal{A}}_3. \end{array}$$

(5.4) Corollary (Igusa [I]). The Schottky locus $\mathcal{RS}_4$ is irreducible, hence is precisely $\mathcal{RJ}_4$. In particular,

$$\mathcal{S}_4^{(\text{big})} = \mathcal{S}_4 = \mathcal{J}_4.$$

Here is a sketch of the proof of (5.3). First, it is clear that the two copies of $K(J(C))$ are indeed in $\beta^{-1}(\alpha(J(C)))$. Consider the equivalence relation $\sim$ on $\overline{\mathcal{RA}}_4^t$ generated by the relation

"$(P_0,\nu_0) \approx (P_1,\nu_1)$ if they are related via the theta symmetry (5.1)".

One verifies that the equivalence class of $(C,\mu) \in \mathcal{RJ}_4$ consists of two copies of the Kummer $K(P(C,\mu))$, as in the statement of the theorem. We end up with a quotient map of β,

$$\overline{\mathcal{RA}}_4^t / \sim \longrightarrow \mathbb{P}^7 / G_3(2,4) \quad ,$$

and by a degeneration argument we conclude that this map is an isomorphism over the image of α, so $\beta^{-1}(\alpha(J(C)))$ cannot contain anything new.

I would like to point out that the fiber of β over a point of $\mathbb{P}^7/G_3(2,4)$ not in image(α) is not known. It is a deformation of the singular variety (consisting of two Kummers meeting along a surface) which is the fiber over a point of image(α), but it should be interesting to have an explicit description.

(5.5) Theorem The compactified map

$$\beta : \overline{\mathcal{RA}}_5^t \longrightarrow \mathbb{P}^{15} / G_4(2,4)$$

is generically finite of degree 119. Its Galois group is (contained in) $SO_8^-(2)$, the special orthogonal group preserving a quadric of Witt-defect 1 in $\mathbb{P}^7(\mathbb{F}_2)$.

(5.6) Theorem The closed Schottky locus $\overline{\mathcal{RS}}_5^t$ has four components:

$$\overline{\mathcal{RS}}_5^t = \overline{\mathcal{RC}^0}^t \cup \overline{\mathcal{RJ}}_5^t \cup \partial^I \overline{\mathcal{RA}}_5^t \cup (\overline{\mathcal{A}}_4^t \times \overline{\mathcal{RA}}_1^t).$$

(I.e. Conjecture (2.11) is true in genus ≤ 5.)

(5.7) Corollary $\mathcal{S}_5 = \mathcal{J}_5$.

The proofs, at present, are very complicated. They rely on detailed knowledge of the structure of the Prym maps

$$\mathcal{P}_g : \overline{\mathcal{RM}}_g \longrightarrow \overline{\mathcal{A}}_{g-1}$$

for $g \lesssim 6$. This knowledge is obtained by applying the tetragonal construction to everything in sight. For instance, $\mathcal{P}_6$ is generically finite of degree 27 [DS] with Galois group

$$WE_6 \approx SO_6^-(2),$$

the symmetry group of lines on a cubic surface: two lines intersect or not according as the corresponding curves are obtained from each other by one tetragonal move or a sequence of two such moves. We can thus define an equivalence relation on $\overline{\mathcal{RA}}_5$ by the theta symmetry (5.1), as in (5.3), but now we get a generically finite relation. Starting with any point of $\overline{\mathcal{RA}}_5$ we get, in the first generation,

$$54 = 27 \cdot 2$$

equivalent objects. Theorem (5.5) involves showing that the second (and last) generation adds another 64 objects fitting together in a highly symmetric configuration, and that this equivalence spans the fibers of β. Theorem (5.6) then requires computation of the local degree of β on each of the 4 known components (these degrees are 1,54,64,0 respectively), and checking the normal bundles to make sure that no extra components arise via blowup (i.e. contribute 0 to the degree).

REFERENCES

[ACGH] E. ARBARELLO, M. CORNALBA, P. GRIFFITHS , J. HARRIS, Geometry of algebraic curves, vol. I, Springer-Verlag, New York (1985).

[AdC1] E. ARBARELLO, C. DE CONCINI, On a set of equations characterizing Riemann matrices, Ann. of Math. 120 (1984), 119-140.

[AdC2] E. ARBARELLO, C. DE CONCINI, Another proof of a conjecture of S.P. Novikov on periods and abelian integrals on Riemann surfaces, Duke Math. J. 54 (1987), 163-178.

[AM] A. ANDREOTTI, A. MAYER, On period relations for abelian integrals on algebraic curves, Ann. Scuola Norm. Sup. Pisa 21 (1967), 189-238.

[AMRT] A. ASH, D. MUMFORD, M. RAPOPORT , Y. TAI, Smooth compactification of locally symmetric varieties, Math. Sci. Press, Brookline (1985).

[B] A. BEAUVILLE, Prym varieties and the Schottky problem. Inv. Math. 41 (1977), 149-196.

[BD] A. BEAUVILLE, O. DEBARRE, Une relation entre deux approaches du probleme de Schottky, preprint.

[Bo] A. BOREL, Some metric properties of arithmetic quotients of symmetric spaces and an extension theorem, J. Diff. Geo. 6(1972), 543-560.

[CG] H. CLEMENS, P. GRIFFITHS, The intermediate Jacobian of the cubic threefold. Ann. Math. 95 (1972), 281-356.

[Co] A. COLLINO, A cheap proof of the irrationality of most cubic threefolds. Bolletino U.M.I. (5) 16-B (1979), 451-465.

[D1] R. DONAGI, The tetragonal construction. AMS Bull. 4 (1981), 181-185.

[D2] R. DONAGI, Big Schottky, Inv. Math. 89 (1987), 569-599.

[D3] R. DONAGI, Non-Jacobians in the Schottky loci, Ann. of Math. 126 (1987), 193-217.

[DS] R. DONAGI, R. SMITH, The structure of the Prym map, Acta Math. 146 (1981), 25-102.

[Du] B.A. DUBROVIN, Theta functions and non-linear equations, Russian Math. Surveys 36, n°2 (1981), 11-92.

[F] J. FAY, Theta functions on Riemann Surfaces, Springer-Verlag, Berlin and New York, 1973, Lecture Notes, Vol. 352.

[FR] H. FARKAS , H. RAUCH, Period relations of Schottky type on Riemann surfaces, Ann. of Math. 92 (1970), 434-461.

[G] M. GREEN, Quadrics of rank four in the ideal of a canonical curve. Invent. Math. 75 (1984), 85-104.

[I] J. IGUSA, On the irreducibility of Schottky's divisor, Journal of the Fac. of Science Tokyo, 28 (1981), 531-545.

[M1] D. MUMFORD, Theta characteristics on an algebraic curve, Ann. Sci. E.N.S. 4 (1971), 181-192.

[M2] D. MUMFORD, Prym varieties I. Contributions to Analysis, 325-350, New York, Acad. Press, 1974.

[M3] D. MUMFORD, Curves and their Jacobians, Ann Arbor, University of Michigan Press, 1975.

[Mu] M. MULASE, Cohomological structure in soliton equations and Jacobian varieties, J. Diff. Geo. 19 (1984), 403-430.

[R] S. RECILLAS, Jacobians of curves with a g_4^1 are Prym varieties of trigonal curves, Bol. Soc. Math. Mexicana 19 (1974), 9-13.

[S] F. SCHOTTKY, Zur Theorie der Abelschen Funktionen von vier Variablen, J. Reine und Angew. Math. 102 (1888), 304-352.

[SJ] F. SCHOTTKY, H. JUNG, Neue Satze über symmetralfunctionen und die Abelschen funktionen, S.-B. Berlin Akad. Wiss. (1909).

[Sh] T. SHIOTA, Characterization of Jacobian varieties in terms of soliton equations. Inv. Math. 83 (1986), 333-382.

[T] A. TJURIN, Five lectures on three dimensional varities, Russ. Math. Surv. 27 (1972).

[Tx] M. TEIXIDOR, For which Jacobi varieties is Sing θ reducible, preprint.

[vG1] B. VAN GEEMEN, Siegel modular forms vanishing on the moduli space of curves, Inv. Math. 78 (1984), 329-349.

[vG2] B. VAN GEEMEN, The Schottky problem and moduli spaces of Kummer varieties, U. of Utrecht thesis, 1985.

[vGvdG] B. VAN GEEMEN, G. VAN DER GEER, Kummer varieties and the moduli space of curves, Am. J. of Math. 108 (1986), 615-642.

[W] W. WIRTINGER, Untersuchungen über Thetafunctionen, Teubner, Berlin, 1895.

[We] G. WELTERS, The surface $C - C$ on Jacobi varieties and 2nd order theta functions, preprint.

[We2] G. WELTERS, A criterion for Jacobi varieties, Annals of Math. 120 (1984), 497-504.

The Cohomology of the Moduli Space of Curves

John L. Harer

The purpose of these notes is to give an exposition of recent work of several people on the topology and geometry of the moduli space of curves. Moduli space may be approached in many different ways. For $g > 1$, it is simultaneously the space of isometry classes of hyperbolic metrics on a surface of genus g, the space of conformal equivalence classes of Riemann surfaces of genus g and the space of algebraic curves of genus g up to isomorphism. This means that there is an elaborate interplay between hyperbolic geometry, complex analysis and algebraic geometry going on. The mapping class group acts properly discontinuously on Teichmüller space with quotient moduli space, so the rational cohomology of moduli space may be identified with that of the mapping class group. This adds a topological and algebraic perspective to things.

The main emphasis in these notes will be on this topological side and we will discuss primarily work of our own. We will, however, spend some time on analysis as we discuss work of Scott Wolpert on the Weil-Petersson geometry of Moduli space. Our main theme then will be the question of how much of the topology and formal geometry of a symmetric space can be found for Teichmüller space and how many of the properties of an arithmetic group can be found for the mapping class group. This will lead us through a discussion of work by Charney and Lee, Harris, Miller, Morita, Mumford, Thurston, Zagier and many others.

We would like to thank Scott Wolpert for a great deal of help with these notes.

The author wishes to thank the C.I.M.E. foundation and the N.S.F. for support of this work.

TABLE OF CONTENTS

Chapter 1. Introduction

Let F be a closed, oriented surface of genus g. The primary object we will be looking at in these notes is the space which parametrizes all the conformal structures carried by the surface F; it is called the moduli space of Riemann surfaces and is denoted M_g. This space can also be defined as the space of hyperbolic metrics on F or the space of algebraic curves of genus g, up to appropriate notions of equivalence. We will not use this third point of view here, but we will constantly be switching between the other two.

§1. First Definitions

We begin with the conformal point of view. Define a marked Riemann surface to be a pair $(R,[f])$ where R is a Riemann surface (= complex 1-manifold), $f: R \to F$ is a homeomorphism and $[f]$ denotes the homotopy class of f. Two marked Riemann surfaces $(R_1,[f_1])$ and $(R_2,[f_2])$ are called equivalent if there is a conformal homeomorphism $h: R_1 \to R_2$ such that $[f_2 \circ h] = [f_1]$. The collection of equivalence classes is denoted T_g; it has a natural topology and is called the Teichmüller space of genus g. (We will discuss this topology later when we introduce Fenchel-Nielsen coordinates.)

The moduli space is obtained by forgetting the marking $[f]$. To make this precise we introduce the mapping class group Γ_g; it is the group of homotopy classes (or, equivalently, isotopy classes) of orientation preserving homeomorphisms of F. The formula

$$[g]\cdot(R,[f]) = (R,[gf])$$

defines an action of Γ_g on T_g; the quotient space is denoted M_g and is called the moduli space of conformal structures on F.

The second way to define T_g and M_g is using hyperbolic geometry $(g>1)$. By a hyperbolic surface we will mean a smooth surface X equipped with a complete Riemannian metric of constant curvature -1. A marked hyperbolic surface is a pair $(X,[f])$ where X is hyperbolic and $f: X \to F$ is a homeomorphism. We say that $(X_1,[f_1])$ is equivalent to $(X_2,[f_2])$ if there is an isometry $h: X_1 \to X_2$ with $[f_2 \circ h] = [f_1]$, and we denote the collection of equivalence classes by the same letter T_g. The justification for this is the uniformization theorem which states that every Riemann surface R is conformally equivalent to one which admits a hyperbolic metric, and this metric is uniquely determined up to isometry by the conformal equivalence class of R.

More generally, one defines the spaces T_g^s, M_g^s and the group Γ_g^s

as follows. Fix s distinct points, ordered $p_1,\ldots,p_s$ on the surface F_g and consider triples $(R,(q_1,\ldots,q_s),[f])$ where R is a Riemann surface, $q_1,\ldots,q_s$ are distinct, ordered points on R and $f:R\to F$ is a homeomorphism with $f(q_i) = p_i$ for each i. In this case $[f]$ denotes the homotopy class of f $\mathrm{rel}\{q_i\}$. The definition of equivalence is the same as before: $(R_1,(q_1^1,\ldots,q_s^1),[f_1]) \sim (R_2,(q_1^2,\ldots,q_s^2),[f_2])$ if and only if there exists a conformal homeomorphism $h:R_1\to R_2$ with $h(q_i^1) = q_i^2$ for each i such that $[f_2\circ h] = [f_1]$. The space of equivalence classes is denoted T_g^s and is again called Teichmüller space.

The mapping class group Γ_g^s is the group of all orientation preserving homeomorphisms $\phi:F\to F$ such that $\phi(p_i) = p_i$ for all i, up to isotopies fixing each p_i. It acts on T_g^s as before and the quotient is moduli space M_g^s.

To define T_g^s using hyperbolic surfaces we remove the s points; set $F_g^s = F_g - \{p_1,\ldots,p_s\}$ and assume $\chi(F_g^s) < 0$. Consider pairs $(X,[f])$ where X is complete hyperbolic of finite area and $f:X\to F_g^s$ is a homeomorphism. Since X is complete and finite area its structure near a puncture is modeled on the pseudosphere. The definition of the equivalence is exactly as in the case $s = 0$.

§2. Fenchel-Nielsen Coordinates

To understand T_g^s better it is necessary to introduce Fenchel-Nielsen coordinates. These will be defined using the description of T_g^s as hyperbolic metrics.

The starting point is the observation that a right hexagon in the hyperbolic plane is determined up to isometry by the lengths of three alternating sides, and these lengths may be chosen to be arbitrary positive numbers. It also makes sense to allow these lengths to go to 0 or ∞; for example, if ℓ_1 goes to 0 in Figure 1.1 we obtain ideal right pentagons, for which there are two parameters, and if ℓ_1 and ℓ_2 both go to 0 we obtain ideal right quadralaterals with one parameter.

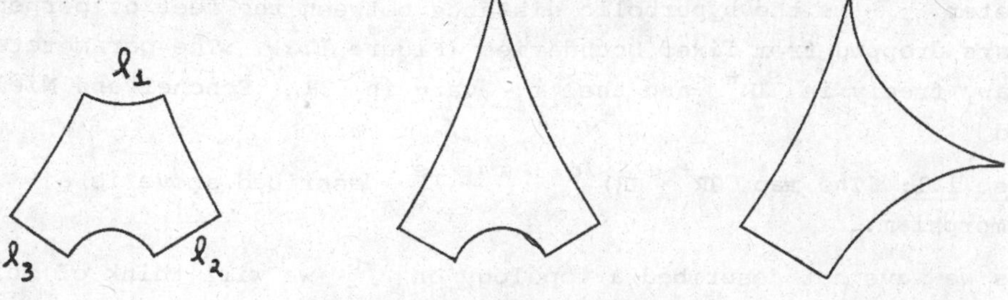

Figure 1.1

Given such a hexagon, form its double across the remaining sides to obtain the basic building block P, a pair of pants (Figure 1.2) with goedesic boundary. The metric on P is now determined by the lengths of its three boundary components and these also are arbitrary. Allowing one or two of these to have length 0 we have the ideal pairs of pants of Figure 1.2.

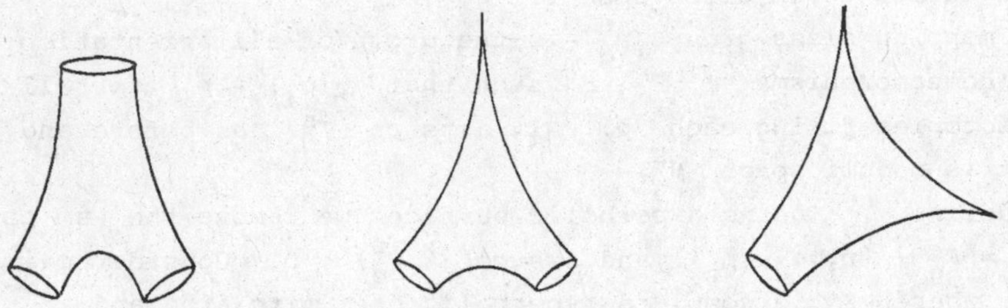

Figure 1.2

Next, fix a partition on F_g^s; this is a collection $C_1, \ldots, C_{3g-3+s}$ of disjoint simple closed curves such that $F_g^s - \{C_i\}$ is the disjoint union of pairs of pants, punctured annuli and twice punctured disks (if $g = 0$ assume $s > 3$, T_0^3 is a single point). We may build a marked hyperbolic surface by glueing hyperbolic pairs of pants (and ideal pairs of pants) together according to the pattern determined by $\{C_i\}$. The Fenchel-Nielsen coordinates are the free parameters for this construction; there are two for each C_i. The first is ℓ_i, the length of C_i; two pairs of pants may be metrically glued along boundary curves to obtain a hyperbolic surface as long as these curves have the same length. The second, the twist parameter τ_i, measures the displacement of the boundary curves along which we glue. The parameter τ_i is the hyperbolic distance between the feet of perpendiculars dropped from fixed boundaries (Figure 1.3). The parameters ℓ_i vary freely in $\mathbb{R}^+$ and the τ_i vary in $\mathbb{R}$. Fenchel and Nielsen proved

Theorem 1.1: The map $(\mathbb{R}^+ \times \mathbb{R})^{3g-3+s} \to T_g^s$ described above is a homeomorphism.

As we have not described a topology on T_g^s we will think of this

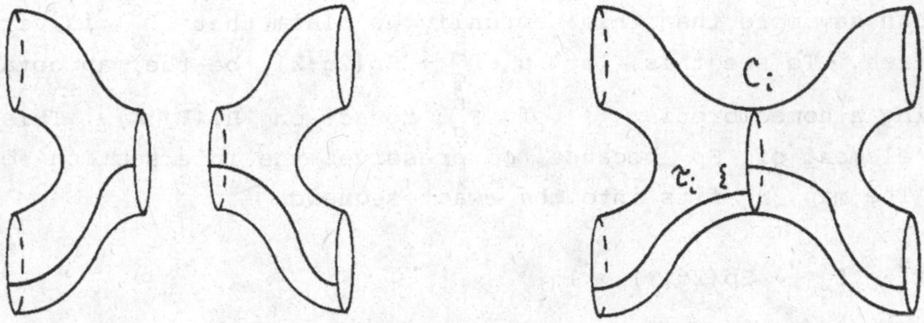

Figure 1.3

result as defining the topology. There is still content, however, to Theorem 1.1 since it is true independent of the choice of partition. Furthermore, we now know that T_g^s is a cell.

The Fenchel-Nielsen coordinates may also be used to describe the Deligne-Mumford compactification $\overline{M}_g^s$ of M_g^s as follows. If $\{C_i\}$ is a partition of F_g^s, allowing some of the ℓ_i to be 0 we obtain a Riemann surface with nodes at those C_i; $\overline{M}_g^s$ is obtained from M_g^s by adjoining these singular surfaces. The complement then has irreducible components $D_0, D_1, \ldots, D_{[g/2]}$ of real codimension 2, where D_0 is the collection of surfaces with a node at a nonseparating curve (and perhaps other nodes) and D_i, $i > 0$, consists of the surfaces with a node at a curve which separates F into surfaces of genus i and g - i.

§3. Homology of M_g^s and Γ_g^s

A well-known result about hyperbolic surfaces is that their length spectrum (the collection of real numbers which occur as lengths of closed geodesics) is discrete. It is not difficult to use this to prove that the action of Γ_g^s on T_g^s is properly discontinuous; i.e. for every compact set $K \subset T_g^s$ the collection of $\phi \in \Gamma_g^s$ such that $\phi(K) \cap K \neq \emptyset$ is finite. This means that M_g^s is a V-manifold or orbifold: each point has a neighborhood modeled on $\mathbb{R}^N$ modulo a finite group. Furthermore, M_g^s is a "rational $K(\Gamma_g^s,1)$"; that is,

$$H_*(M_g^s;\mathbb{Q}) \cong H_*(\Gamma_g^s;\mathbb{Q}).$$

One can say more than this; actually we claim that Γ_g^s is virtually torsion free. To see this, let $\mu : \Gamma_g^s \to Sp(2g;\mathbb{Z})$ be the map obtained by allowing a homeomorphism ϕ of F_g^s to act on $H_1(F_g;\mathbb{Z})$. This gives an element of Sp because ϕ preserves the intersection form on F_g. The map μ fits into the exact sequence

$$1 \to T_g^s \to \Gamma_g^s \xrightarrow{\mu} Sp(2g;\mathbb{Z}) \to 1 \qquad (S_0)$$

where T_g^s is the Torelli group. Now look at G_n, the full congruence subgroup of level n in $Sp(2g;\mathbb{Z})$ which is defined as the subgroup of matrices congruent to the identity mod n. For $n \geq 3$, G_n is torsion free and it is also well-known that T_g^s is torsion free. Therefore, the congruence subgroup $\Gamma_g^s[n] = \mu^{-1}(G_n)$ will also be torsion free, $n \geq 3$. Its index is the order of the finite group $Sp(2g;\mathbb{Z}/n\mathbb{Z})$ so the claim is established.

The quotient $\mathcal{T}_g^s/\Gamma_g^s[n] = \mathcal{M}_g^s[n]$ is called the moduli space of curves with <u>level</u> <u>n</u> <u>structure</u>. It is a manifold and we have

$$H_*(\Gamma_g^s[n];\mathbb{Z}) \cong H_*(\mathcal{M}_g^s[n];\mathbb{Z}).$$

At times it will be necessary to compare the homology groups of Γ_g^s as we vary g and s. For s we have the exact sequence

$$1 \to \pi_1(F_g^s) \to \Gamma_g^{s+1} \xrightarrow{\eta} \Gamma_g^s \to 1, \qquad (S_1)$$

defined as follows. Let η be the map obtained by forgetting p_{s+1}. If ϕ lies in $\mathrm{Ker}(\eta)$, then ϕ is isotopic to the identity fixing $p_1,\ldots,p_s$. Following p_{s+1} under this isotopy determines an element of $\pi_1(F_g^s)$; the sequence S_1 is derived from this. The Lyndon-Hockshield-Serre spectral sequence may then be used to relate $H_*(\Gamma_g^s)$ to $H_*(\Gamma_g^{s+1})$. When we vary g, however, there is no natural way of mapping Γ_g^s to $\Gamma_{g+1}^{s'}$ so it becomes necessary to introduce mapping class groups of surfaces with boundary. In Chapter 6 we will describe these and show how to use them to prove that for $g \gg k$ $H_k(\Gamma_g^s)$ is independent of g.

§4. Algebraic Structure of the Mapping Class Group

Because $H_*(\mathcal{M}_g^s)$ and $H_*(\Gamma_g^s)$ are so intimately related, we will need to know some facts about the algebraic structure of Γ_g^s.

We first discuss a finite presentation for Γ_g^s. Let $C \subset F_g^s$ be a simple closed curve; the <u>Dehn twist</u> of C, denoted τ_C, is (the isotopy class of) the homeomorphism of F_g^s obtained by splitting along C, rotating one side 360° to the right and reglueing (Figure 1.4).

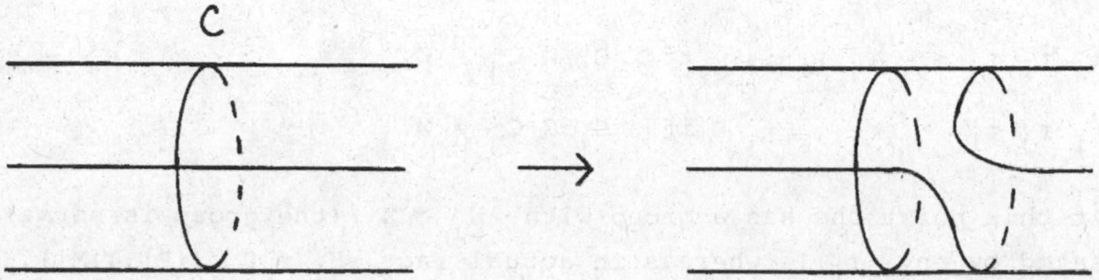

Figure 1.4

Dehn proved that Γ_g^s is generated by Dehn twists on a finite number of curves ([D]) and Humphries determined the minimal number of twist generators necessary (the $2g+1$ curves of Figure 1.5 when $s = 0$).

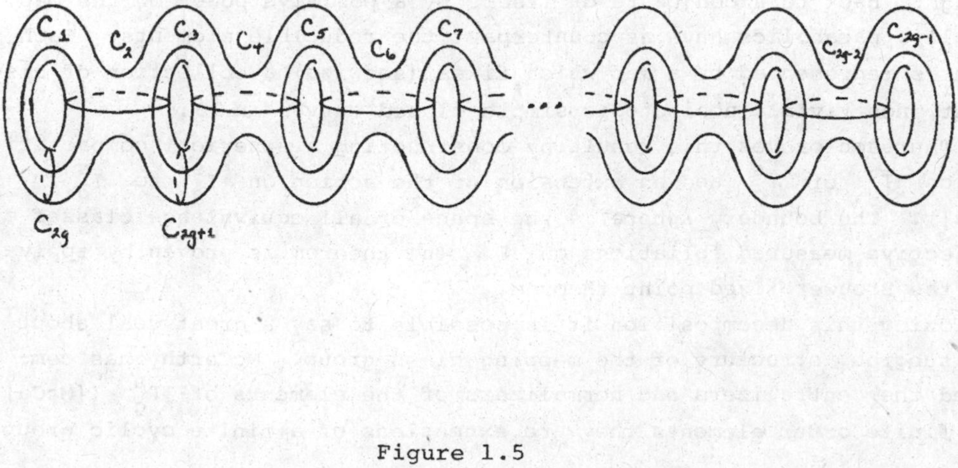

Figure 1.5

McCool([Mc]) gave an indirect proof that Γ_g is finitely presented, but the first explicit presentation was provided by Hatcher and Thurston ([HT]). This was simplified by Wajnryb ([Wa]) whose presentation uses as generators $\tau_i = \tau_{C_i}$, $1 \le i \le 2g+1$, and has $\binom{2g+1}{2} + 3$ relations $(g > 2)$. The first of these are the braid relations:

$$\tau_i \tau_j = \tau_j \tau_i \quad \text{if} \quad C_i \cap C_j = \emptyset,$$

$$\tau_i \tau_j \tau_i = \tau_j \tau_i \tau_j \quad \text{if} \quad C_i \cap C_j \neq \emptyset.$$

At this point one has a group with $H_1 \cong \mathbf{Z}$ (the group is normally generated by any τ_i), whereas in actual fact $H_1 = 0$ ([P], [M]). The "lantern relation" (see [Hl]) is added next, giving a group with $H_1 = 0$. Now the result has $H_2 = 0$ whereas H_2 should be $\mathbf{Z}$([Hl]). To fix this we add the "Chinese lantern relation" to get $H_2 \cong \mathbf{Z}$. It turns out that this is a presentation of the mapping class group of a surface with 1 boundary component. To get Γ_g^0 we add one more relation, called the "boundary relation". We refer the reader to [Wa] for explicit forms of these relations. Presentations for the groups Γ_g^s can easily be obtained using (S_1).

Next we briefly describe Thurston's classification of the elements of Γ_g^s. This classification is modeled on the decomposition of $PSL_2\mathbb{R}$ into elliptic, hyperbolic and parabolic elements. The analogue of the elliptics are the elements of finite order in Γ_g^s; each such may be realized as an isometry of some hyperbolic metric. Corresponding to hyperbolics are the pseudo-Anosovs; these are represented by maps which preserve a pair of transverse, measured foliations (with singularities) and they are distinguished by the fact that no element of $\pi_1(F)$ is brought back to a conjugate of itself by a positive power of the map. Finally, parabolics have as counterpart the reducible elements. Each such is represented by a map which fixes (setwise) a collection of disjoint, nontrivial, nonisotopic simple closed curves in F.

Thurston proves this result by constructing a spherical compactification $\bar{T}_g^s$ of T_g^s and an extension of the action on Γ_g^s to $\bar{T}_g^s$ ([T3]). The boundary sphere is the space of all equivalence classes of projective measured foliations on F. The theorem is proven by applying the Brouwer fixed point theorem.

Using this decomposition it is possible to say a great deal about the subgroup structure of the mapping class group. McCarthy has computed the centralizers and normalizers of the elements of Γ_g^s ([McCa]). For finite order elements they are extensions of a finite cyclic group by a mapping class group, for pseudo-Anosovs they are finite-by-infinite cyclic and for reducibles they are a mixture of the two. Long ([Ll]) showed that if $H < \Gamma_g$ is finitely generated and contains a free group of rank 2 generated by 2 pseudo-Anosovs (for example, Γ_g itself), then H contains uncountably many maximal subgroups of infinite index and the Frattini subgroup of H, which is the intersection of all maximal subgroups of H, is a torsion group. Finally, Birman, Lubotsky and McCarthy showed that every solvable subgroup of Γ_g is virtually abelian ([BLM]).

All of these properties are similar to those of discrete subgroups of linear algebraic groups. In fact, slightly weaker forms of them would follow immediately if we had a discrete, faithful linear representation of Γ_g. Thus we are led naturally to the question: is Γ_g actually linear? Or even more: is Γ_g arithmetic?

§5. The Analogy with Symmetric Spaces

The problem which will motivate us in this notes is this: how close is Teichmüller space to being a symmetric space and how close is Γ_g^s to being arithmetic? When G is a linear algebraic group defined over $\mathbb{Q}$ and $\Gamma < G_{\mathbb{Q}}$ is arithmetic (see Chapter 4 for definitions), then Γ acts properly discontinuously on the symmetric space $X = G/K$, K maximal compact in G. The space X is diffeomorphic to Euclidean space and the quotient $\Gamma\backslash X$ is a V-manifold; it follows that $H_*(\Gamma;\mathbb{Q}) \cong H_*(\Gamma\backslash X;\mathbb{Q})$. This suggests an analogy between T_g^s and X and between Γ_g^s and Γ and most of what we will do springs from this analogy.

The first question which then arises is whether there exists some G such that $T_g^s = G/K$ and $\Gamma_g^s = \Gamma$. Ivanov was the first to announce a proof that this is not the case; in fact he shows that Γ_g^s is not arithmetic in any linear algebraic group. In Chapter 4 we will give a proof of this due to Bill Goldman.

Even though Γ_g^s is not arithmetic, we can still ask which properties it shares with the arithmetic groups. This is the theme of Chapters 2, 3, 4, 6 and 8. Among other things we will see that Teichmüller space admits a Borel-Serre bordification (Chapter 3) and that the mapping class group is a virtual duality group (Chapter 4), satisfies homological stability as g goes to infinity (Chapter 6) and admits a formula for its Euler-characteristic which involves the Riemann zeta-function (Chapter 8). All of these are properties of arithmetic groups.

We can also ask how much of the formal geometry of a symmetric space may be found for T_g^s. There are two well-known metrics for T_g^s. The first is the Teichmüller metric; it is only a Finsler metric and the geometry it provides is quite distorted. In fact, Royden ([R]) showed that in this metric the group of isometries of T_g^s is exactly the mapping class group, so the situation is very unlike that of a symmetric space. The second is the Weil-Petersson metric. This metric is Kähler and is much more suited to our purposes. It has strictly negative Ricci curvature and holomorphic sectional curvatures. In Chapter 5 we will discuss Scott Wolpert's striking work on the symplectic and Hermitian geometry this metric gives for Teichmüller space. This geometry is intimately tied to the complex structure on; in fact, in 3 we will outline how Wolpert uses the Weil-Petersson Kähler form to give an analytic proof that $\overline{M}_g^s$ is projective.

Chapter 2: Triangulating Teichmüller Space

The purpose of this chapter is to describe an ideal triangulation of Teichmüller space which is compatible with the action of the mapping class group. The construction works for any $s \geq 1$ but we will restrict to the case where $s = 1$ to keep the exposition simple. The original idea for this triangulation is due to Thurston and uses hyperbolic geometry; the details for this approach were provided by Bowditch and Epstein and will be given in §3. The first complete proof, however, was given by Mumford using the conformal point of view and was based on results of Strebel. (Yet another proof was provided more recently by Epstein and Penner using the interpretation of Teichmüller space as conjugacy classes of discrete, faithful representations of the fundamental group of a surface in $SO(2,1)$). We will present Mumford's proof first, in §2, after giving the combinatorial structure of the triangulation in §1.

§1 The Simplicial Complex

Let F be a closed, oriented surface of genus $g \geq 1$ and let $*$ be a basepoint in F. The isotopy class (rel $*$) of a family $\alpha_0,\ldots,\alpha_k$ of simple closed curves in F through $*$ will be called a rank-k arc-system if α_i intersects α_j only at $*$ when $i \neq j$ and the family satisfies the nontriviality condition that no α_i is null-homotopic and no distinct α_i and α_j are homotopic (rel $*$). The maximum rank an arc-system can have is $6g-4$ since $6g-3$ curves will decompose F into triangles so that no more curves can be added without violating nontriviality.

i) Definition of A

Form a simplicial complex A by taking a k-simplex $\langle\alpha_0,\ldots,\alpha_k\rangle$ for each rank-k arc-system in F and identifying $\langle\beta_0,\ldots,\beta_\ell\rangle$ as a face of $\langle\alpha_0,\ldots,\alpha_k\rangle$ if $\{\beta_i\} \subset \{\alpha_j\}$. By the remarks above, A has dimension $6g-4$. Points in A correspond to pairs (α,w) where α is an arc-system represented by curves $\alpha_0,\ldots,\alpha_k$ and w is a collection of non-negative weights $w_0,\ldots,w_k$ on the α_i such that $w_0+\ldots+w_k = 1$.

ii) Definition of A_∞

A family of curves is said to fill the surface F if each component of its complement is simply connected. Define A_∞ to be the subcomplex of A consisting of all simplices $\langle\alpha_0,\ldots,\alpha_k\rangle$ such that $\{\alpha_i\}$ does not fill F. There is a natural action of Γ_g^1 on A given by $[f]\cdot\langle\alpha_0,\ldots,\alpha_k\rangle = \langle f(\alpha_0),\ldots,f(\alpha_k)\rangle$ and since this action preserves A_∞ it restricts to an action on $A - A_\infty$. The main theorem of this chapter is:

Theorem 2.1: There is a homeomorphism $\omega\colon T_g^1 \to A - A_\infty$ which commutes with the action of the mapping class group Γ_g^1.

Sections 2 and 3 of this chapter are devoted to the proof of this theorem. Before we go on, however, we will discuss in some detail the case where $g = 1$.

iii) Example, $g = 1$

A single simple closed curve cannot fill the torus, but any arc-system with 2 or more curves (rank = 1 or 2) must do so. This means that A_∞ contains only the vertices of A; these in turn may be identified with $\mathbb{Q} \cup \{\infty\}$: if $\{m,\ell\}$ is a basis for $\pi_1 F$ corresponding to two non-homotopic simple closed curves meeting only at $*$, then any other simple closed curve α through $*$ represents $a_1 m + a_2 \ell$ in $\pi_1 F$ with a_1 prime to a_2. Associating a_1/a_2 to α gives the bijection between A_∞ and $\mathbb{Q} \cup \{\alpha\}$. It is easy to see that two curves with parameters (a_1,a_2) and (b_1,b_2) are isotopic to ones which meet only at $*$ exactly when $a_1b_2 - a_2b_1 = \pm 1$. We may therefore identify A with $\mathbb{H} \cup \mathbb{Q} \cup \{\infty\}$, where $\mathbb{H}$ is the hyperbolic plane as in figure 2.1 (upper half plane model) or figure 2.2 (Poincare model).

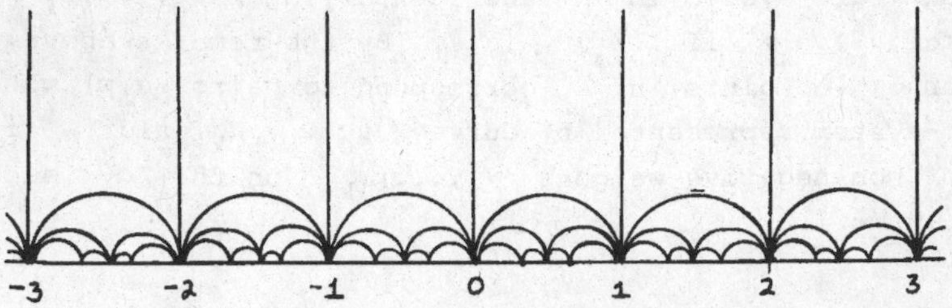

Figure 2.1

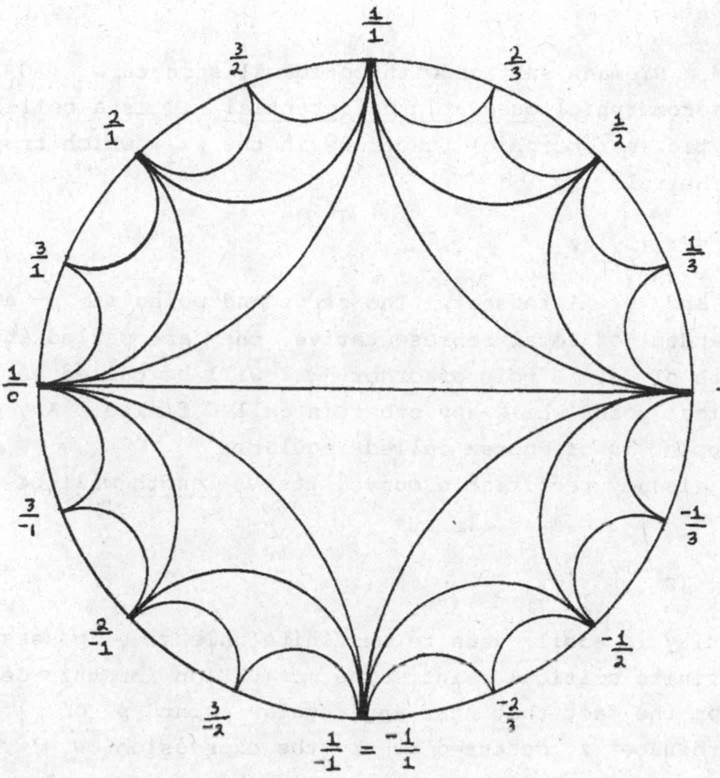

Figure 2.2

Since T_1^1 may be identified with $\mathbb{H}$, the picture gives an illustration of theorem 2.1.

It should be pointed out that the situation is much more complicated in higher genus. Since it takes $2g$ curves to fill the surface, A_∞ contains the $2g - 2$ skelton of A; however, it also contains pieces of the skeleta of A up to codimension 2. The existence of these higher dimensional cells will turn out to be a red herring, however, because we will see in chapter 4 that A_∞ has the homotopy type of a wedge of spheres of dimension $2g - 2$.

§2 The Conformal Point of View: Strebel Quadratic Differentials.

In this section we will give the Mumford-Strebel proof of Theorem 2.1. The main ingredient is the theory of quadratic differentials.

i) Definitions

Let R be a Riemann surface with conformal structure $\{(U_i, z_i)\}$. An analytic (meromorphic) quadratic differential ϕ is a collection $\{\phi_i\}$ of analytic (meromorphic) functions in the z_i which transform according to the rule

$$\phi_i(z_i)dz_i^2 = \phi_j(z_j)dz_j^2$$

whenever U_i and U_j intersect. The zeros and poles of ϕ are clearly independent of local representative, they are called the critical points of ϕ. A pole of order ≥ 2 will be called an infinite critical point while any other is called finite. Any non-critical point is of course called regular.

If $\gamma \subset R$ is any rectifiable curve, its ϕ-length will be defined to be

$$|\gamma|_\phi = \int_\gamma |\phi(z)|^{1/2} |dz|$$

and this quantity is easily seen to be finite unless γ passes through an infinite critical point. The motivation for this definition comes from the fact that near any regular point p of ϕ with conformal coordinate z centered at p the expression $w = \int \sqrt{\phi(z)}dz$ makes sense and gives a new parameter w in which ϕ is identically equal to 1. Then $|\gamma|_\phi = \int_{\gamma'} |dw|$ is the ordinary Euclidean length of γ' where γ' is the image of γ in the w-plane.

With the metric defined by $||_\phi$ it is possible to talk about geodesics in R. The two types important to us are the horizontal and vertical trajectories of ϕ: A smooth curve γ is called horizontal if $\arg \phi(z)dz = 0$ along γ and vertical if $\arg \phi(z) = \pi$ along γ. The horizontal trajectories of ϕ are the (unique) maximal horizontal curves through every regular point of ϕ, with a similar definition for vertical trajectories. These trajectories give two perpendicular foliations of R - critical points ; if we add the critical points (by taking the closure of the leaves) we obtain singular foliations F_h and F_v of R.

ii) Horocyclic Quadratic Differentials

Suppose ϕ is a quadratic differential on R with exactly one pole of order 2 at $p \in R$ and no other poles. Suppose further that all the horizontal trajectories of ϕ which consist only of regular points are closed curves. In a neighborhood of the pole p there is a distinguished parameter ζ so that ϕ has the representation $\frac{c\,d\zeta^2}{\zeta^2}$. When $c < 0$ we will call such a quadratic differential horocyclic (this terminology will make sense after §3). The trajectory structure of ϕ near a zero is an n-pronged singularity, $n \geq 3$, illustrated in figure 2.3 for $n = 3, 4$. Around the pole the structure is as shown in figure 2.4. The horizontal trajectories are concentric circles around p, while the vertical ones are wheel-spokes emanating from p.

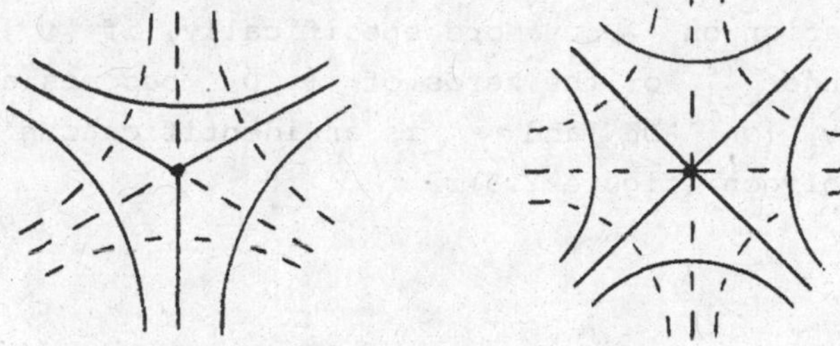

n = 3 n = 4

Figure 2.3

horizontal vertical

Figure 2.4

The reason we are interested in quadratic differentials is the following result of Strebel [S].

Theorem 2.2: Let R be a closed Riemann surface of genus g and p a point of R. Then there exists a horocyclic quadratic differential ϕ on R with its pole at p. The differential ϕ is unique up to multiplication by positive scalars.

Strebel proves this theorem by solving the following extremal mapping problem. Let z be a coordinate at p and consider the family of all conformal embeddings $\lambda: D\rho \to R$ where D_ρ is a disk of radius ρ in the w plane, $\lambda(0) = p$ and $\left|\frac{dz}{dw}(0)\right| = 1$. Using a normal families argument he shows that a λ exists which maximizes ρ and it is unique up to multiplication by a constant. The inverse of λ is then the distinguished parameter for a horocyclis quadratic differential ϕ whose nonsingular horizontal trajectories are the image under λ of the circles w = constant.

The map λ has the added property that it extends to a map of the closed disk $\bar{\lambda}: \bar{D}_\rho \to R$, exhibiting R as $\bar{D}\rho/\sim$ where $\sim$ is an identification on $\partial\bar{D}\rho$. More specifically, if $\{v_i\}$ is the inverse image under $\bar{\lambda}$ of the zeros of ϕ, $\bar{D}\rho$ becomes a polygon with vertices v_i on $\partial\bar{D}\rho$ and $\sim$ is an identification of the edges of this polygon (figure 2.5).

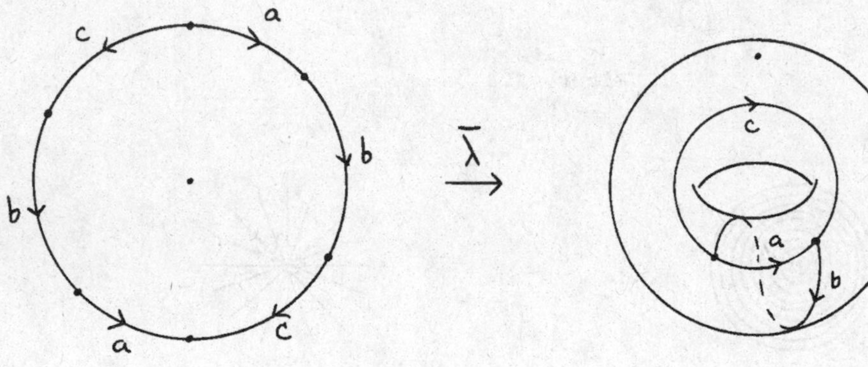

Figure 2.5

iii) Definition of $\omega: T_g^1 \to A - A_\infty$

We are finally ready to define the map $\omega: T_g^1 \to A - A_\infty$. Let R be a Riemann surface, $p \in R$, and let $f: (R,p) \to (F,*)$ represent a marking of R; the triple $(R,p,[f])$ determines a point of T_g^1. By applying Theorem 2.2 we obtain a horocyclic quadratic differential on R centered at p. Let $q_1,\ldots,q_n$ be the zeros of ϕ and let $\gamma_1,\ldots,\gamma_m$ be the singular leaves of the horizontal foliation F_h determined by ϕ. The γ_i are closed horizontal arcs whose interiors consist of regular points and whose endpoints lie in $\{q_j\}$. Each nonsingular leaf of the vertical foliation is a loop based at p which is perpendicular to F_h and meets exactly one of the γ_i. Two such are homotopic (rel p) in $R - \{q_i\}$ if and only if they intersect the same γ_i; select one for each γ_i and call it $\overline{\alpha}_i$. Let $\alpha_i = f(\overline{\alpha}_i)$; the collection $\{\alpha_i\}$ is then an arc-system in F. (An example is given in figure 2.6.) If γ_i has ϕ-length ℓ_i and ℓ is the sum of the ℓ_i, we let $w_i = \ell_i/\ell$ to obtain positive weights on the α_i. The map ω is now defined by

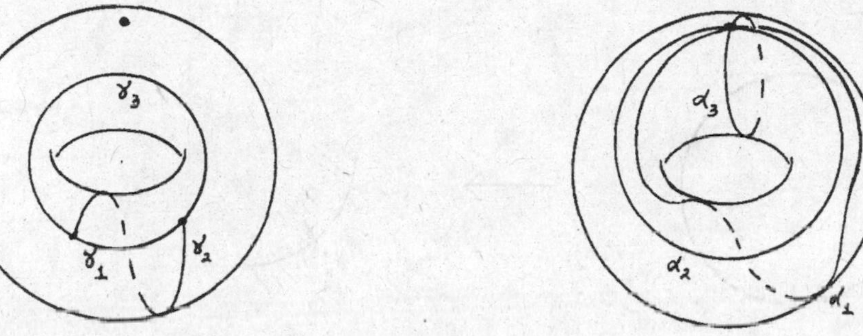

Figure 2.6

$$\omega((R,p,[f])) = (\{\alpha_i\}, \{w_i\}).$$

iv) Definition of $\eta = \omega^{-1}: A - A_\infty \to T_g^1$

To define the inverse η of ω, let $(\{\alpha_i\}, \{w_i\})$ represent a point of $A - A_\infty$ where the arc-system $\{\alpha_i\}$ has rank $k - 1$. The complement of the dual graph Ω to $\{\alpha_i\}$ is a 2-disk. Split F along Ω to obtain a polygon P with 2k-sides, an identification $\sim$ of the edges of P and a surjective map $f_0: P \to F$ taking the center point $*$ of P to p, the boundary of P onto Ω and commuting with $\sim$ so that it induces a homeomorphism $P/\sim \to F$. Let the edges of P have labels γ_i^+, γ_i^- where γ_i^+ is paired to γ_i^- by $\sim$ and $f_0(\gamma_i^+) = f_0(\gamma_i^-)$ is the edge of Ω which meets α_i. We will now use the combinatorial data $(P, \{\gamma_i^\pm\}, \{w_i\})$ to build a marked Riemann surface.

Begin with the closed unit disk $D = \{z \in \mathbb{C}: \|z\| \leq 1\}$ and choose a homeomorphism $f_1: P \to D$ taking $*$ to 0 such that $f_1(\gamma_i^+)$ and $f_1(\gamma_i^-)$ have Euclidean length πw_i for each i. The edges γ_i^+ and γ_i^- map to arcs in ∂D which we denote with the same symbols. Also let $\{v_i\}$ be the image under f_1 of the vertices of P. By identifying each γ_i^+ with γ_i^- via the composition of the inversion $z \to 1/z$ with a rotation we obtain a Riemann surface R_0 with singularities $\{q_j\} = f_2(\{v_i\})$ where $f_2: D \to R_0$ is the quotient map (see figure 2.7).

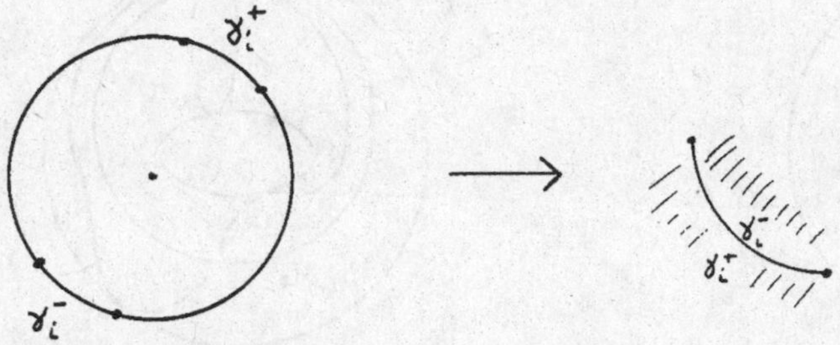

Figure 2.7

The total angle at each q_j is $n_j\pi$ where n_j is the number of v_i which map to q_j. Since these angles are each commensurable with 2π, a standard argument using branched coverings extends the complex structure to the q_j and gives a closed Riemann surface R.

To define the marking $f: R \to F$ we merely complete the diagram

$$\begin{array}{ccc} D & \xleftarrow{\ f_1\ } & P \\ f_2 \downarrow & & \downarrow f_0 \\ R_0 \cong R & \longrightarrow & F \end{array}$$

this can be done because the quotient maps f_0 and f_2 are compatible.

The expression $\phi = \frac{-dz^2}{z^2}$ defines a quadratic differential on D with a double pole at the origin such that the ϕ-lengths of γ_i^+ and γ_i^- are their Euclidean lengths πw_i. This means ϕ is compatible with the identifications and therefore descends to R. The data which ϕ determines is clearly the original weighted arc-system $(\{\alpha_i\},\{w_i\})$, so the map

$$\eta((\{\alpha_i\},\{w_i\})) = (R,p,[f])$$

is the inverse of ω as required.

§3 The Hyperbolic Point of View

Because of the importance of theorem 2.1 we will present an alternate proof of it in this section. The original idea for this proof is due to Thurston and was explored by Mosher in [Mo]; the details were worked out by Bowditch and Epstein in [BE]. Throughout this section Teichmüller space will be treated as the space of marked hyperbolic structures (complete, finite area) on a surface of genus g with 1 puncture.

i) Definition of ω

First we define the map $\omega: T_g^1 \to A-A_\infty$. Let X be a complete hyperbolic surface of finite area with one puncture and let $f: X \to F - \{*\}$ be a homeomorphism; $(X,[f])$ represents a point of T_g^1. Begin with an embedded horocycle C_0 around the puncture and let X_0 be X with the open punctured disk enclosed by C_0 removed. The function $\rho; X_0 \to \{t\geq 0\}$ which associates to a point of X_0 its minimum distance from C_0 has level set C_t at time t; C_t is called the quasi-horocycle at distance t. For small t C_t is smooth but in general C_t has singularities at those points of X_0 which are equidistant from 2 or more points of C_0. These singularities form a 1-dimensional connected graph $\Omega \subset X$ whose complement is a once-punctured disk (compare Ω in §2 and see figure 2.8).

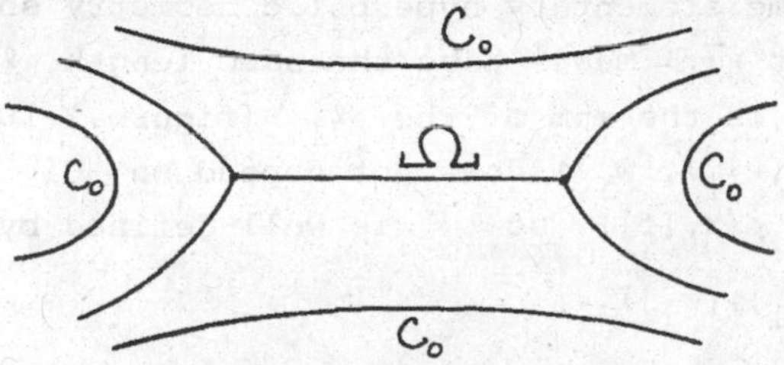

Figure 2.8

The construction of the map ω is now similar to the construction in §2: the geodesics perpendicular to C_0 form a non-smooth singular foliation F_v of X_0, where each nonsingular leaf of F_v is the union of two geodesic segments joining Ω to C_0 (figure 2.9). For each edge γ_i of Ω there is a

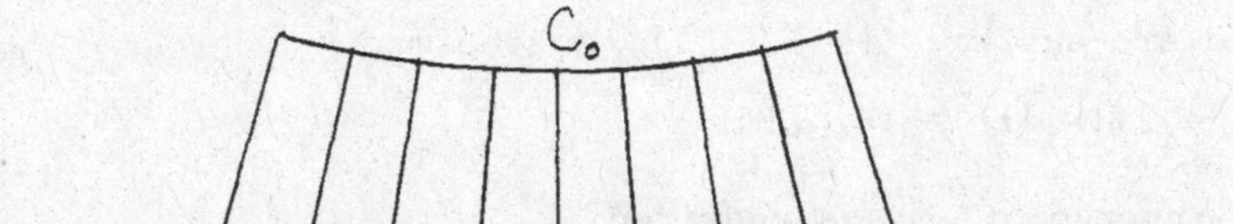

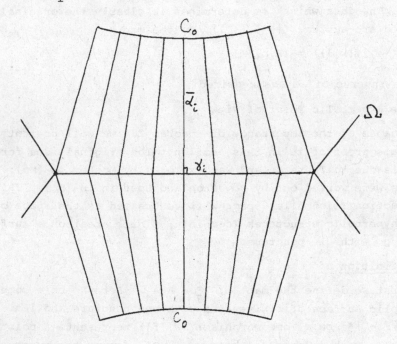

Figure 2.9

unique leaf $\bar{\alpha}_i$ of F_v perpendicular to γ_i; $\bar{\alpha}_i$ is a smooth geogesic in X_0 and has a completion to a bi-infinite geodesic in X. Apply the map f and add on the point * to get an embedded arc α_i in F based at *; then $\{\alpha_i\}$ is the arc-system associated to $(X,[f])$. To determine the weights w_i look at the leaves of F_v which meet the singular points of Ω. The intersection of these leaves with C_0 divides C_0 into segments each of which meets exactly one $\bar{\alpha}_i$. Some elementary hyperbolic geometry shows that the two segments that $\bar{\alpha}_i$ meets have the same length ℓ_i; set $w_i = \ell_i/\ell$ where ℓ is the sum of the ℓ_i (figure 2.10). It is easy to check that $(\{\ell_i\},\{w_i\})$ does not depend on C_0 or the equivalence class of $(X,[f])$, so ω is well-defined by the formula

$$\omega((X,[f])) = (\{\alpha_i\},\{w_i\}).$$

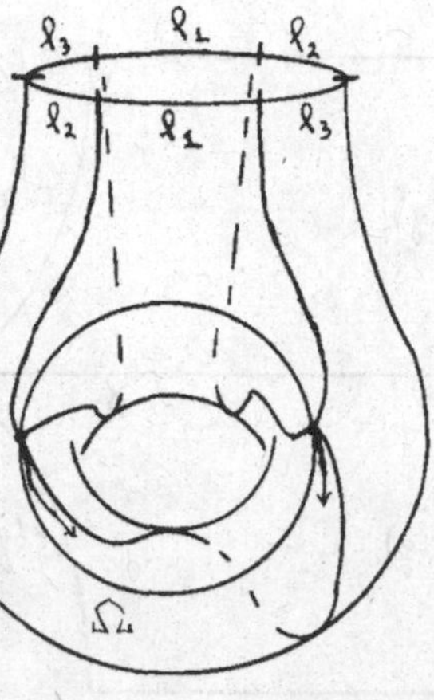

Figure 2.10

ii) Definition of $\eta = \omega^{-1}: A-A_\infty \to T_g^1$

The inverse map $\eta : A-A_\infty \to T_g^1$ is defined by constructing the strips between the singular leaves of the foliation F_v explicitly (figure 2.10), and then glueing them together to build X_0. The problem will be to match the angles around the singular points of F_v (= the vertices of Ω).

We begin with a model for the combinatiorial structure of X_0, namely take $F_0 = F -$(small open disk around *), $C_0 = \partial F_0$ and let $\overline{\alpha}_i = \alpha_i \cap F_0$. Also take the dual graph Ω with edges γ_i and vertices q_j and add to the picture embedded edges e_j^k from q_j to C_0, one for each homotopy class of paths from q_j to C_0 in $F_0 - \{\overline{\alpha}_i\}$ (these correspond to the singular leaves of the foliation F_v). Splitting along $\{e_j^k\}$ divides F_0 into strips, each containing a single pair α_i, γ_i.

It is clear that strips as illustrated in figure 2.11 exist in the hyperbolic plane with geodesic sides and horocyclic tops and bottoms (curvature $\equiv 1$). We choose $\{w_i\}$ very small, but in the correct projective class, and restrict to those strips which have length w_i on top and bottom and are symmetric with respect to reflection through γ (this was the case for the strips on X between the singular leaves of F_v). There is then a 2-parameter family of such strips where the

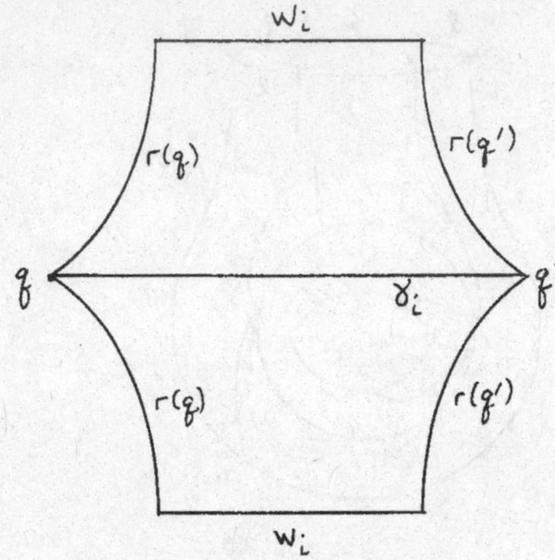

Figure 2.11

parameters are the distances $r(q)$ and $r(q')$ from the endpoints q, q' of γ_i to the top (or bottom). To glue the strips back together we only need to have constructed them so that the lengths of the edges e_j^k from the vertex q_j to C_0 are the same in each strip. By symmetry this distance can only depend on q_j, so any collection of v positive numbers $r_1,\ldots,r_v (r_i = r(q_i)$ and $v =$ the number of vertices of Ω) gives a singular hyperbolic structure on F_0 with horocyclic boundry and with singularities at the vertices of Ω. The total angle around q_i will be denoted Θ_i; the surface constructed above will be nonsingular when each $\Theta_i = 2\pi$. The problem is to show this occurs for an appropriate choice of $r_1,\ldots,r_v$.

Define $\mu : (0,\infty)^v \to (0,\infty)^v$ by setting $\mu(r_1,\ldots,r_v) = (\Theta_1,\ldots,\Theta_v)$. If q_i has valence n_i in Ω, then since $n_i \geq 3$ the point $P = (2\pi,\ldots,2\pi)$ lies well in the interior of the convex hull H of the 2^v points $(\varepsilon_1 n_1\pi,\ldots,\varepsilon_v n_v\pi)$, where ε_i equals 0 or 1. We claim that if the w_j are chosen small enough, then $\mathrm{im}(\mu)$ will contain enough of H to contain P. In fact, as the $w_i \to 0$, $\mathrm{im}(\mu)$ will converge to an open set containing all of the interior of H. To see this, look at the upper portion (which is isometric to the lower portion) of a strip as in Figure 2.11. Label the top edge λ_i and let $\psi(q)$, $\psi(q')$ be the interior angles at q, q' respectively. There is a bound to how small $r(q)$ and $r(q')$ can be before γ_i becomes tangent to λ_i; however, as we shorten λ_i this bound goes to 0. Therefore if we let $r(q)$ and $r(q')$ be near their minimum values, as w_i goes to 0 $\psi(q)$ and $\psi(q')$ approach $\pi/2$.

If follows that $(n_1\pi,\ldots,n_v\pi)$ lies in the limit of the closure of $\mathrm{im}(\mu)$ as all w_i goes to 0. Now suppose all the r_i are short; pick $i_1,\ldots,i_s$ and keep $r_{i_1},\ldots,r_{i_s}$ short while the others are allowed to vary. It is clear that as $r_j \to \infty$, $\Theta_j \to 0$. Furthermore, if $r(q)$ is kept short while $r(q')$ grows long in our strip, then $\psi(q)$ increases to a maximum value greater than $\pi/2$. From this it is not hard to see that the wall of $H: (t_1 n_1\pi,\ldots,t_v n_v\pi)$ where $t_{i_1} = \ldots = t_{i_s} = 1$ and all other t_j vary between 0 and 1, lies in the closure of $\mathrm{im}(\mu)$ as the w_i go to 0. On the other hand, any point in the dual wall where $t_{i_1} = \ldots = t_{i_s} = 0$ can be reached by letting $r_{i_1},\ldots,r_{i_s}$ go to ∞.

By now the picture is clear. Bowditch and Epstein complete the argument by showing that μ is proper and $1-1$. The intermediate value theorem then guarantees that for w_i small enough, $(2\pi,\ldots,2\pi)$ lies in $\mathrm{im}(\mu)$.

Chapter 3: The Borel-Serre Bordification of Teichmüller Space

In this chapter we begin the study of properties shared by Teichmüller space and Symmetric spaces by proving the existence of a Borel-Serre bordification for T_g^s.

Let G be a linear algebraic group defined over $\mathbb{Q}$ and let X be the symmetric space G/K where K is maximal compact in G. In this situation Borel and Serre ([BS]) construct a manifold-with-corners $\overline{X}$ with $\overline{X} - \partial\overline{X} = X$ such that if $\Gamma \subset G_{\mathbb{Q}}$ is any arithmetic group, then the action of Γ on X extends to a properly discontinuous action of Γ on $\overline{X}$ with $\Gamma\backslash\overline{X}$ compact. The manifold $\Gamma\backslash\overline{X}$ is then used to make cohomological computations for Γ.

Our goal will be the proof of:

Theorem 3.1: There exists a piecewise-linear manifold (or a smooth manifold with corners) W which is contractible and has interior T_g^s such that the action of the mapping class group Γ_g^s on T_g^s extends to a properly discontinuous action of Γ_g^s on W with W/Γ_g^s compact.

Notice that W/Γ_g^s is then a compactification of moduli space M_g^s. In chapter 4 we will use W to study the cohomology of Γ_g^s and M_g^s.

The first Borel-Serre bordification of T_g^s was discovered by Harvey ([Har]) who constructed W by adding on copies of lower genus Teichmüller spaces (crossed with Euclidean spaces of the appropriate dimension) at infinity. We will give a description of Harvey's bordification in §2; however, it will be constructed inside Teichmüller space, not externally as Harvey did originally. This way it will be easy to see how Γ_g^s act on W; the only difficult part will be identifying T_g^s with the interior of W.

Before we give Harvey's construction, however, we will use the triangulation of chapter 2 to find another description of W. The advantage of this second point of view is that W is constructed combinatorially and this will allow us to analyze the homotopy type of ∂W (chapter 4, Theorem 4.1). A second advantage of the combinatorial construction of W is that when $s > 0$ W retracts Γ_g^s-equivariantly onto a spine $Y \subset W$ of dimension $4g-3$. This implies:

Theorem 3.2: The moduli space M_g^s has the homotopy type of a finite cell-complex of dimension $4g-4+s$, $s > 0$. In particular,

$H_k(M_g^s) = 0, \quad k > 4g-4+s, \quad s>0$ and

$H_k(M_g;\mathbb{Q}) = 0, \quad k > 4g-5.$

It would be nice if we had an equivariant spine $Y \subset T_g$ of dimension $4g-5$ so that we could remove the $\mathbb{Q}$-coefficients in the theorem. Thurston has constructed a geometric candidate for Y and shown how to retract T_g onto it; but there is no combinatorial description of Y available that would allow us to decide if it has the best possible dimension. We will describe this spine in §3.

§1. The Combinatorial Construction

In this section we give the first construction of the Borel-Serre bordification W of Teichmüller space. Just as in chapter 2 we will only consider the case $s = 1$ to keep the exposition simple (see [H3] for the general case). Recall that we have identified $A-A_\infty$ with T_g^1, we will work directly with A in building W and Y.

i) Definition of W and Y

Let A^o be the first barycentric subdivision of A. The complex A^o has a vertex of <u>weight</u> k for each rank-k arc-system $\langle\alpha_o,\ldots,\alpha_k\rangle$ in F and an r-cell for each chain of $r+1$ inclusions of arc systems. The subcomplex A_∞^o is defined similarly; we set Y^o equal to the union of all the simplices of A^o which have no face in A_∞^o. (By simplex we always mean closed simplex unless stated otherwise.) The complex Y^o is a spine for Teichmüller space, we will see shortly that it has dimension $4g-3$. Let A^{oo} and A_∞^{oo} be the second barycentric subdivisions of A and A_∞ respectively. We define W to be the collection of all simplices of A^{oo} which have no face in A_∞^{oo}. W is a regular neighborhood of Y^o and the group Γ_g^1 acts on the pair (W,Y^o) with both W/Γ_g^1 and Y^o/Γ_g^1 compact.

ii) Description of Y

First we will study the spine Y^o. The key fact is that Y^o is the first barycentric subdivision of the dual complex Y of A. This allows us to describe Y directly: Y has a k-cell for each rank $6g-4-k$ arc-system which fills F and the cell corresponding to $\{\alpha_i\}$ is a face of the cell corresponding to $\{\beta_j\}$ if $\{\beta_j\} \subset \{\alpha_i\}$. It is instructive to enumerate the low dimensional cells of Y: a 0-cell of Y corresponds to a maximal arc-system in F, maximality means that the curves of the arc-system triangulate F. A 1-cell of Y corresponds to an arc-system consisting of $6g-4$ curves; these curves cut F into 4g-4 triangles and 1 square. The 1-cell is attached to the two 0-cells which correspond to the two possible completions of the arc-system to a maximal one (figure 3.1). A 2-cell of Y corresponds to an arc-system with 6g-5 curves which cut F into either 4g-5 triangles and 1 pentagon, or 4g-6 triangles and 2 squares. The corresponding 2-cells of Y are illustrated in figure 3.2. This process continues until we reach the

Figure 3.1

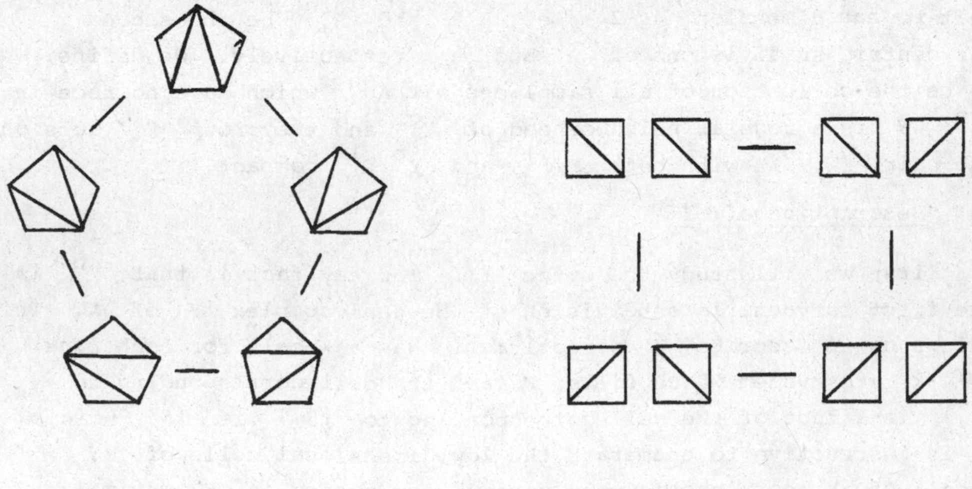

Figure 3.2

case of an arc-system consisting of $2g$ curves, since this is the minimum number of curves mecessary to fill F. In this situation the curves cut F into a single $4g$-gon and the possible completions to higher rank arc-systems describe cells which fit together to form a $4g-4$ sphere. Y has a $4g-3$ cell attached along this $4g-4$ sphere.

As a subset of T_g^1, Y corresponds to those surfaces R with basepoint p such that the graph Ω given by the Strebel differential ϕ has "enough" edges of maximal ϕ-length. By this we mean that if ℓ is the largest length of any edge of Ω, then we may collapse the edges of Ω whose length is less than ℓ without changing the topological type of R. The dimension of the (open) cell of Y which contains $(R,p,[f])$ is $6g-3$ minus the number of maximal length edges of Ω.

The genus 1 case is illustrated in figure 3.3; the vertices of Y

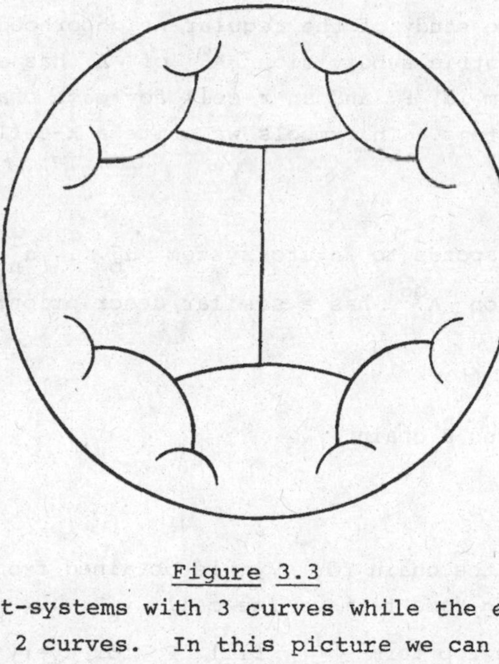

Figure 3.3

correspond to cut-systems with 3 curves while the edges correspond to cut-systems with 2 curves. In this picture we can see why the vertices of A have no dual 2-cells: the link of a vertex is PL-homeomorphic to $\mathbb{R}$, not to S^1 as it would have to be. We can also see how to retract T_1^1 onto Y as follows. The complement of Y in T_1^1 has one component for each point of A_∞ and T_1^1 acts transitively on these components. Furthermore, every point in $T_1^1 - Y$ lies on a unique geodesic ray which begins on Y and tends to a point of A_∞.

Flowing in along these geodesics rays collapses T_1^1 onto Y; by choosing the flow in one component of $T_1^1 - Y$ and using the action of Γ_1^1 to extend it to all of T_1^1 it may be made Γ_1^1-equivariant. This means the flow descends to moduli space M_1^1; here M_1^1 is homeomorphic to $\mathbb{R}^2$, but is collapsed onto Y/Γ_1^1 which is an interval.

The general case is directly analogous to the case where $g = 1$. The dimension of the complex Y is $4g-3$ because the cells in A_∞^o have links which are contracible, not spherical, so they have no dual cells. The Teichmüller space T_g^1 can be Γ_g^1-equivariantly retracted onto Y by "flowing" along straight lines (in the simplicial structure provided by A) away from A_∞ onto Y. The construction of these simplicial flow lines is somewhat technical and will be omitted (the construction requires a proof that the entire complex A is contractible; see [H3]).

iii) <u>Description of W</u>

Now we pass to the study of the regular neighborhood W. Recall that the first barycentric subdivision A^o of A has a vertex for each rank k arc-system in F and an r-cell for each chain of $r + 1$ inclusions of arc-systems. In symbols we write a k-cell of A^o as

$$\beta_o \subset \beta_1 \subset \ldots \subset \beta_k$$

where each β_i corresponds to an arc-system $\alpha_o^i, \ldots, \alpha_{n_i}^i$. The second barycentric subdivision A^{oo} has a similar description; its k-cells are written

$$\gamma_o < \gamma_1 < \ldots < \gamma_k$$

where each γ_i denotes a chain

$$\beta_o^i \subset \ldots \subset \beta_{m_i}^i$$

and $\gamma_i < \gamma_{i+1}$ means the chain for γ_i is obtained from the chain for γ_{i+1} by omitting some terms. The cell $\gamma_o < \ldots < \gamma_k$ lies in W if and only if the top term $\beta_{m_i}^i$ fill F for every i and it lies in ∂W when, in addition, the bottom term β_o^i does not fill F for every i. Using this description it is not hard now to check that W is a PL manifold with boundary by analyzing the links of the cells in $W-\partial W$ and in ∂W. For details of this in the general case we refer to reader to [H3]; here we will only deal with the case $g = 1$ where we can actually draw W.

A picture of W is given in figure 3.4. To keep the illustration form getting too cluttered we have drawn only the outline of W and the spine Y. To see the full

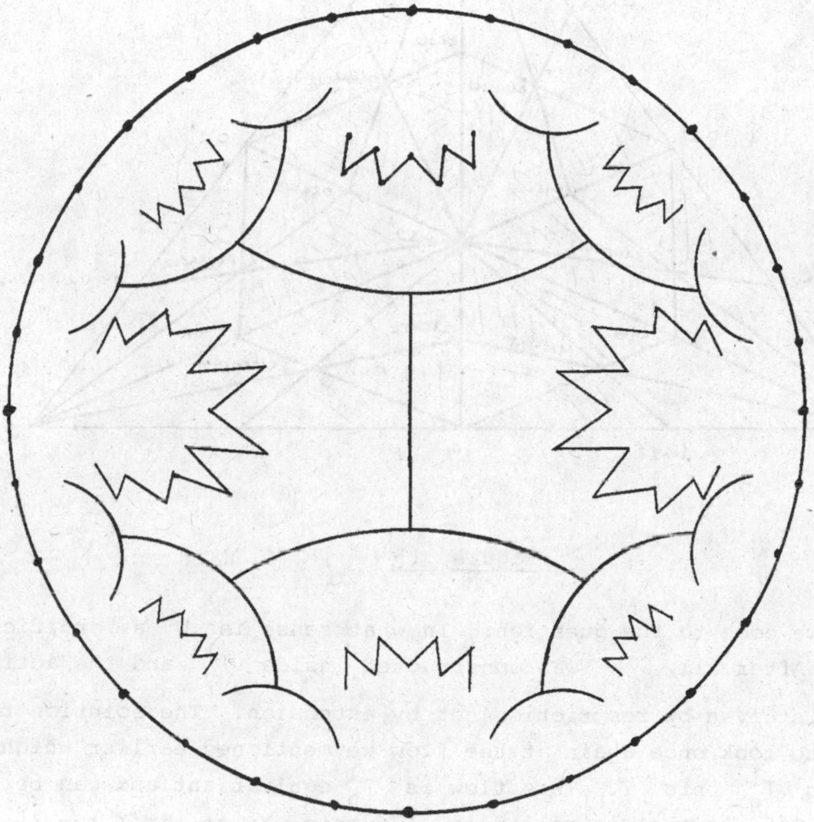

Figure 3.4

picture we have pulled out a simplex of A in figure 3.5. The vertices of the simplex are labeled 0, 1, 2, corresponding to three curves α_o, α_1, α_2 of a maximal arc-system and the vertices of A^{oo} have been labeled using the notation introduced above. The part of W contained in the simplex is shaded.

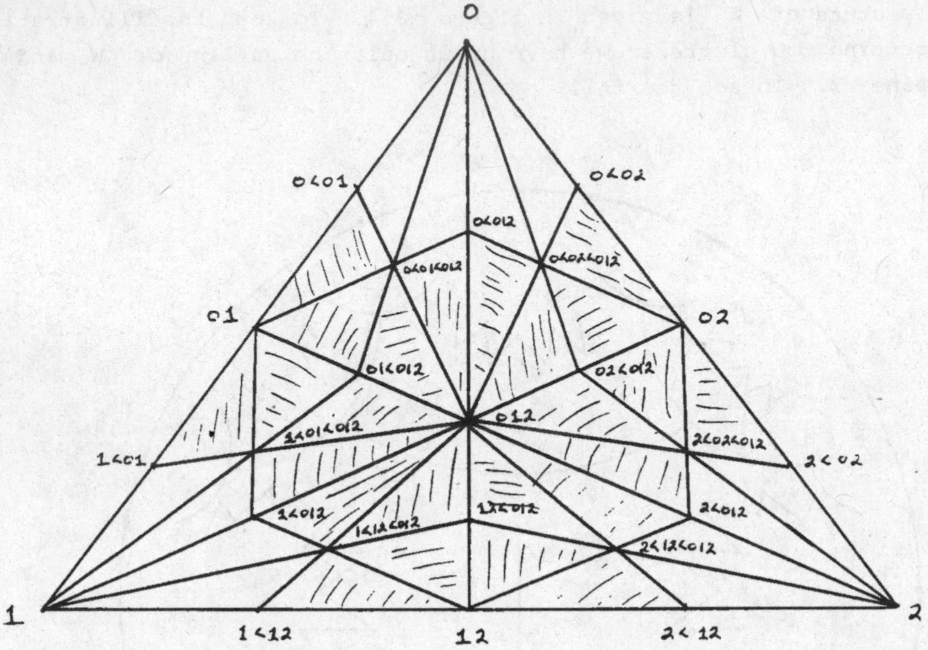

Figure 3.5

Now we come to the question: In what sense is W a bordification of T_g^1? After all, W was constructed inside T_g^1 and the action of Γ_g^1 is given by restriction not by extension. The solution to this is to look once again at the flow we mentioned earlier which collapses T_g^1 onto Y. This flow is Γ_g^1-equivariant and can be parameterized so that it collapses T_g^1 onto Y at time $t = 1$ and provides a homeomorphism of T_g^1 into itself at any time $t < 1$. After some rescaling by linear coordinate changes in the simplices of A^o we can arrange that the flow at time $t = 1/2$ gives a homeomorphism of T_g^1 onto $W - \partial W$.

If we remove Y and add on A_∞ the inverse of the flow is well-defined and retracts everything onto A_∞. This provides the necessary tool to prove

Lemma 3.3: ∂W is Γ_g^1-equivariantly homotopy equivalent to A_∞.

We omit the details of the proof (see[H3]). This lemma will be used in chapter 4 to prove that the groups Γ_g^s are virtual duality groups.

§2 The Geometric Construction

In this section we will give a geometric construction of Harvey's bordification of Teichmüller space. This construction has the advantage over the one in §1 that it works for any value $s \geq 0$ but the disadvantage that it does not provide a spine of the correct dimension (compare §3). While these two constructions do not give the exact same object, they are nevertheless equivalent in the sense that for $s > 0$ there is a map from the combinatorial bordification to the geometric bordification which is a Γ_g^s-equivariant homeomorphism on $\mathcal{T}_g^s$ and a homotopy equivalence on the boundary.

i) <u>Geodesic Length Functions</u>

Let F be a surface of genus g with s punctures and let (X,f) represent a point of $\mathcal{T}_g^s$ where X is complete hyperbolic of finite area and $f: X \to F$ is a homeomorphism. For any simple closed curve $C \subset F$ there is a unique closed geodesic $\gamma \subset X$ such that $f(\gamma)$ is freely homotopic to C; γ will be simple. We define the <u>geodesic length function</u> $\ell_C: \mathcal{T}_g^s \to \mathbb{R}^+$ by associating to (X,f) the length of γ. It is a standard result that the function ℓ_C is C^∞ for every C.

Before we define W we state (without proof) an elementary result from hyperbolic geometry.

<u>Lemma 3.4</u>: There exists a number $\varepsilon > 0$ such that if X is any complete hyperbolic surface and γ_1, γ_2 are two simple closed geodesics of length $\leq \varepsilon$ then γ_1 and γ_2 are disjoint.

ii) <u>Definition of W</u>

Fix $\varepsilon > 0$ as in the lemma and set:

$$W = \{(X,[f]) \in \mathcal{T}_g^s : \ell_C((X,[f])) \geq \varepsilon \text{ for every simple closed curve } C \subset F\}$$

It is clear that the action of Γ_g^s on $\mathcal{T}_g^s$ restricts to W where it is properly discontinuous. A theorem of Mumford says that W/Γ_g^s is compact. We summarize the main facts about W.

<u>Proposition 3.5</u>: W is a real analytic manifold-with-corners which admits a properly discontinuous action of Γ_g^s such that W/Γ_g^s is compact. There is a diffeomorphism $W - \partial W \to \mathcal{T}_g^s$ which commutes with the action of Γ_g^s.

The interior of W is $\overset{\circ}{W} = \{(X,[f]) : \ell_C((X,[f])) > \varepsilon$ for every $C\}$ and $W - \overset{\circ}{W} = \partial W$ is $\{(X,[f]) : \ell_C((X,[f])) = \varepsilon$ for at least one $C\}$. It is not hard to see that $\overset{\circ}{W}$ is open in $\mathcal{T}_g^s$:

let $(X,[f]) \in \overset{\circ}{W}$ and suppose $\gamma_1,\ldots,\gamma_t \subset X$ have the same length and have shorter length than any other closed geodesics in X. Since the length spectrum on X is discrete there is a neighborhood of $(X,[f])$ in T_g^s so that if $(X',[f'])$ lies in this neighborhood and has shortest curves $\gamma_1',\ldots,\gamma_n'$ then each $f'(\gamma_i')$ is freely homotopic to some $f(\gamma_j)$. If we put $C_j = f(\gamma_j)$ then the intersection of this neighborhood with $\cap \ell_{C_i}^{-1}((\varepsilon,\infty))$ lies in $\overset{\circ}{W}$.

iii) <u>The Structure of ∂W; Definition of Z_g^s</u>

The first step in analyzing ∂W is to define a new simplicial complex Z_g^s. The isotopy class of a family $\{C_0,\ldots,C_k\}$ of disjoint simple closed curves in F will be called a <u>rank-k</u> <u>curve-system</u> if the curves satisfy the nontriviality condition that no C_i is freely homotopic to a point and no C_i and C_j are freely homotopic to one another $(i \neq j)$. The complex Z_g^s has a k-simplex $\langle C_0,\ldots,C_k\rangle$ is a face of $\langle C_0',\ldots,C_\ell'\rangle$ exactly when $\{C_i\} \subset \{C_j'\}$. Since a curve system can contain at most $3g-3+s$ curves, Z_g^s has dimension $3g-4+s$. We will show in Chapter 4 that for $s = 1$ Z_g^1 is homotopy equivalent to A_∞ (this is also true for $s > 1$, but we have not given the definition of A_∞ in this case; it can be found in [H3]).

The complex Z_g^s will act as parameter space for ∂W: suppose $C = \{C_0,\ldots,C_k\}$ is a rank-k curve-system in F. Define

$$T_C = \{(X,[f]) \in T_g^s\colon \ell_{C_i}((X,[f])) = \varepsilon \text{ for every } i\}.$$

When k is maximal $(=3g-4+s)$, the curves of C form a partition of F so we may use them to define Fenchel-Nielsen coordinates on T_g^s. These coordinates provide an identification of T_C with $\mathbb{R}^{3g-3+s}$ since only the twist parameters can vary. When k is not maximal we use the fact that C can be included in a maximal curve-system to see that T_C is again homeomorphic to Euclidean space. This time the lengths of the extra curves may vary so T_C has dimension $6g-5+2s-k$ (codimension $k+1$).

When the curve-system C has rank 0, the subspace T_C has codimension 1; these subspaces act like walls which cut W out of T_g^s. Not all of the wall T_C will border W; define

$$T_C^+ = \{(X,[f]) \in T_C\colon \ell_{C'} > \varepsilon \text{ whenever } C' \not\subset C\}$$

and

$$\overline{T}_C^+ = \{(X,[f]) \in T_C\colon \ell_{C'} \geq \varepsilon \text{ for every } C'\}.$$

$$\partial W = \bigcup_{C \text{ rank } 0} \overline{T}_C^{\,+} = \coprod_{C} T_C^+$$

By lemma 3.4, if $C = \{C\}$ and $C' = \{C'\}$ are two rank-0 curve-systems then $\overline{T}_C^+$ intersects $T_{C'}^+$ if and only if C and C' are isotopic to disjoint curves, and in that case $\overline{T}_C^+ \cap \overline{T}_{C'}^+ = \overline{T}_{C\cup C'}^+$.

Each of the spaces T_C may be thought of as a kind of Teichmüller space; let $F_C = F - C$ (we treat the new holes as punctures), then T_C may be identified naturally with $\mathbb{R}^{k+1} \times$ (Teichmüller space of F_C) where the first factor corresponds to the twist-parameters on the k+1 curves of C. Using flows like the one we will construct in iv, T_C^+ may be identified with T_C and $\overline{T}_C^+$ with the Borel-Serre bodification of T_C. In particular, this tells us that $\overline{T}_C^+$ is contractible and proves:

Lemma 3.6: ∂W is Γ_g^s-equivariantly homotopy equivalent to the complex Z_g^s.

Our analysis of W is now easy to complete. Each point of T_C^+ has a neighborhood U in T_g^s so that if $(X,[f]) \in U$, then the curves having the shortest length in X lie in C. If we complete C (arbitrarily) to a partition, the resulting Fenchel-Nielen coordinates identify $U \cap W$ with $\mathbb{R}^n \times Q$ where $N = 6g-7+2s-k$ and Q is the upper orthant in $\mathbb{R}^{k+1}$ where $\ell_{C_i} \geq 0$ for every C_i in C. These coordinates give W the required structure of manifold-with-corners (see [BS]).

iv) Deformation of T_g^s onto $W - \partial W$

To finish the proof of Theorem 3.6 we must show how to map T_g^s onto $W - \partial W$ equivariantly. This will be done by constructing a flow on T_g^s which moves the surfaces which have short curves in a direction which increases the lengths of those curves, but fixes the surfaces which do not have short curves. This is easy to do one curve-system at a time using Fenchel-Nielsen coordinates; it is considerably harder to find a Γ_g^s-equivariant flow which simultaneously increases the length of all short curves. The one we describe here is due to Scott Wolpert.

Begin by selecting $\varepsilon > 0$ such that 3ε satisfies the conditions of Lemma 3.4; that is, any two closed geodesics of length $\leq 3\varepsilon$ are disjoint. Choose a C^∞ function $\phi: [0,\infty] \to [0,1]$ such that

$$\phi[0,\varepsilon] \equiv 1, \quad \phi[2\varepsilon,\infty) \equiv 0$$

and ϕ is decreasing on $[\varepsilon, 2\varepsilon]$.

The flow will be constructed using the gradients of the functions ℓ_C. To regulate the flow we need a lemma:

Lemma 3.7: There exist functions $\kappa_C \in C^\infty(T_g)$, one for every isotopy class of simple closed curve in F, such that at each point $(X,[f])$ of T_g^s, the formula

$$\sum_{C:\ell_C \le 3\varepsilon} \kappa_C (\mathrm{grad}\ \ell_C)\ \ell_{C'} = \phi(\ell_{C'})$$

holds for every C' such that $\ell_{C'}(X,[f]) \le 3\varepsilon$.

Proof: Let $C_1,\dots,C_n$ be disjoint simple closed curves in F. Twisting on C_i generates a vector field t_{C_i} on T_g^s (see chapter 5).

The set $\{C_i\}$ can be completed to a partition of F where the t_{C_i} become coordinate vector fields in the resulting Fenchel-Nielsen coordinates; this implies the t_{C_i} are linearly independent.

Define $g(\ ,\ )$ to be the Weil-Peterson metric on T_g^s and let ω be the corresponding 2-form $(\omega(v,w) = g(Jv,w)$ where J is the complex structure on $T_g^s)$. The duality formula (again see chapter 5) of Wolpert says

$$\omega(t_C,\ ) = g(it_C,\ ) = -d\ell_C$$

In particular, this means that grad $\ell_C = it_C$ and tells us that $(\mathrm{grad}\ \ell_{C_i})d\ell_{C_j} = g(\mathrm{grad}\ \ell_{C_i}, \mathrm{grad}\ \ell_{C_j}) = g(it_{C_i}, it_{C_j}) = g(t_{C_i}, t_{C_j})$.

Linearly independence of the t_{C_i} then implies that the matrix $((\mathrm{grad}\ \ell_{C_i})d\ell_{C_i})$ is positive definite and therefore nonsingular.

From this we see immediately that the functions κ_C exist at each point $(X,[f]) \in T_g^s$ such that $\ell_C(X,[f]) \le 3\varepsilon$. It is not hard to check that $\kappa_C = 0$ if $\ell_C(X,[f]) > 2\varepsilon$ so setting $K_C = 0$ where it is not already defined extends κ_C to all of T_g^s. Using the fact that ℓ_C and $g(\ ,\ )$ are C^∞ one proves easily that κ_C is C^∞. □

Now we can define the flow on T_g^s. Let V be the vector field whose value at $(X,[f])$ is given by

$$V = \sum_{C:\ell_C \le 3\varepsilon} \kappa_C (\mathrm{grad}\ \ell_C).$$

The flow ψ_t generated by V is clearly Γ_g^s-invariant; the length functions satisfy $\frac{d\ell_C}{dt} = \phi(\ell_C)$. At time $t = \varepsilon$, ψ_ε is the required differomorphism of T_g^s onto $W - \partial W$. □

§3. Thurston's Spine for T_g^0

The method which we described in §1 for constructing a spine $Y \subset T_g^1$ may be adapted to work for any value of $s \geq 1$. It does not, however, work when $s = 0$ and there is at present no known combinatorial description of a spine for T_g^0. Thurston has given a geometric description of what ought to be the spine in this case and shown how to retract T_g^0 onto it Γ_g^0-equivariantly [T2]. Unfortunately we are unable to say whether it is best possible (lowest dimension) as we could in the earlier case.

The subspace $Y \subset T_g^0$ is easy to describe; it consists of all marked hyperbolic surfaces X which have the property that the shortest closed geodesics $\gamma_1, \dots, \gamma_n$ (length $\gamma_1 = \dots =$ length γ_n and all other closed geodesics are longer) fill the surface X. It is easy to see that two shortest closed geodesics can meet in at most one point so the number n is bounded for fixed g. Notice that when $(X, [f])$ lies in Y all the curves in $\{\gamma_i\}$ must be nonseparating, otherwise they could not fill X. This means that a flow on Teichmüller space which collapeses everything onto Y cannot preserve shortest geodesics; any separating shortest geodesic must end up longer than some non-separating one.

Thurston constructs a flow of T_g^0 onto Y as follows. Let X be a hyperbolic surface with marking f and suppose $\{\gamma_1, \dots, \gamma_n\}$ is a set of simple closed geodesics in X which do not fill X. For later use we set $\Lambda = \{f(\gamma_i)\}$. Choose a simple closed geodesic $\gamma \subset X$ such that each γ_i is either disjoint from γ or equals γ and let X_0 be obtained by splitting X along γ (X_0 may be disconnected). The surface X_0 is naturally included in a complete hyperbolic surface X_1 as a deformation retract (X_1 has infinite area and flares out at ∂X_0). Any geodesic arc α properly imbedded in X_0 extends uniquely to a bi-infinite geodesic in X_1 which is embedded; split X_1 open along this geodesic line and insert a strip from $\mathbb{H}^2$ bounded by two nearby lines. The surface X_0 is replaced by a new surface with the property that any closed geodesic in X_0 which meets α is now longer. If we perform this operation on several arcs every geodesic in X_0 can be lengthened.

By matching the change in the lengths of the two curves of ∂X_0 we may reglue to form a new surface which has the property that any geodesic not meeting γ has been lengthened (as has γ). The infinitesimal version of this construction defines a vector field on all of T_g^0 which we denote V_Λ. For completeness, when Λ is a finite set of simple closed curves which fill F we set $V_\Lambda = 0$.

The next step is to patch these vector fields together with a partition of unity. For any $\varepsilon > 0$ and any Λ as above define U_Λ to be the set of all $(X,[f]) \in T_g^0$ such that

$$\{C: \ell_C((X,[f])) \le L_X + |\Lambda|\varepsilon\}$$

is exactly Λ, where $|\Lambda|$ denotes the cardinality of Λ and L_X is the length of the shortest geodesic in X. For ε small enough the sets U_Λ form a covering of T_g^0: choose B such that for every $(X,[f])$ the number of closed geodesics of length less than $L_X + 1$ is not more than B and let $\varepsilon < 1/B$. Let $\gamma_1,\dots,\gamma_n$ be the geodesics on X of length less than $L + 1$ with $\ell(\gamma_i) \le \ell(\gamma_{i+1})$ for every i. If there is a first $i > 1$ such that $\ell(\gamma_i) > L + \varepsilon i$, set $\Lambda = \{f(\gamma_1),\dots,f(\gamma_{i-1})\}$; otherwise set $\Lambda = \{f(\gamma_1),\dots,f(\gamma_n)\}$. In either case $(X,[f])$ lies in U_Λ.

Now we choose a partition of unity $\{\lambda_\Lambda\}$ subordinate to $\{U_\Lambda\}$ and define a vector field V_ε on T_g^0 by the formula

$$V_\varepsilon = \sum_\Lambda \lambda_\Lambda V_\Lambda.$$

This makes sense because if two sets U_Λ and $U_{\Lambda'}$ intersect, then either $\Lambda \subset \Lambda'$ or $\Lambda' \subset \Lambda$. The flow generated by V_ε is Γ_g^0-equivariant and deforms T_g^0 into

$$Y_\varepsilon = \{(X,[f]): \{C: \ell_C((X,[f])) < L_X + \varepsilon\} \text{ fills } F\}.$$

Letting ε go to zero gives the desired retraction onto Y.

Chapter 4: How Close is the Mapping Class Group to Being Arithmetic?

Let G be an algebraic subgroup of GL_n defined over $\mathbb{Q}$, $G_{\mathbb{Q}}$ the group of $\mathbb{Q}$-points of G and $G_{\mathbb{Z}} = G_{\mathbb{Q}} \cap GL_n\mathbb{Z}$. A subgroup $\Gamma < G_{\mathbb{Q}}$ is called arithmetic if it is commensurable with $G_{\mathbb{Z}}$. The mapping class groups have many properties in common with the arithmetic groups. In the following list (taken from [Se 1]) Γ denotes an arithmetic group or a mapping class group:

(1) Γ is finitely presented,

(2) Γ has only finitely many conjugacy classes of finite subgroups,

(3) Γ is residually finite,

(4) Γ is virtually torsion free,

(5) for any torsion free subgroup $\hat{\Gamma}$ of finite index in Γ there exists a finite complex which is a $K(\hat{\Gamma},1)$,

(6) the virtual cohomological dimension (vcd) of Γ is finite.

References for properties (1)-(6) when Γ is arithmetic may be found in [Sl]. For the mapping class group see [Wa] and [HT[for (1), [Mc] for (3); (4), (5) and (6) may be proven using the complex Y of Chapter 3 (although all three are well-known and follow from more standard results).

In §1 of this chapter we will establish for the mapping class group the next property on Serre's list for arithmetic groups:

(7) $H^q(\Gamma;\mathbb{Z}[\Gamma])$ is zero except for a single value of $q(q = \mathrm{vcd}(\Gamma))$ for which it is a free $\mathbb{Z}$-module I; thus Γ is a virtual duality group as defined by Bieri and Eckmann [Bi Ec].

This is Corollary 4.2 below. When Γ is arithmetic in an algebraic group G which is simple and has $\mathbb{Q}$-rank $r_{\mathbb{Q}} \geq 2$, the following holds:

(8) Every normal subgroup of Γ is either finite index, or is finite and central.

This property fails for Γ_g^s, because the Torelli group T_g^s = Ker $(\Gamma_g^s \to Sp(2g;\mathbb{Z}))$ is normal and is neither finite nor finite index. In §2 we will use this to show Γ_g^s is not arithmetic in G when $r_{\mathbb{Q}}(G) > 1$. Concerning the possibility that Γ_g^s might be arithmetic in G with $r_{\mathbb{Q}}(G) = 1$, we will also see in §2 that vcd Γ_g^s turns out to be the wrong value for this to be true $(g \geq 3)$. Thus we will have proven the result (announced first by Ivanov) that Γ_g^s is not arithmetic for $g \geq 3$. In fact, one extra step will show that Γ_g^s cannot be a lattice (discrete, cofinite volume) in any algebraic group G. The more general question of whether Γ_g^s admits any faithful representation at all in an algebraic group remains open (compare [McCa], [L1] and [L2]).

§1. The Mapping Class Group is a Virtual Duality Group

Let Γ be an arithmetic group in a linear algebraic group G and let X be the symmetric space associated to G. If $\overline{X}$ denotes the Borel-Serre bordification of X, then the action of Γ on X extends to a properly discontinuous action on $\overline{X}$ (in fact the action of $G_{\mathbb{Q}}$ extends to $\overline{X}$). The boundary $\partial\overline{X}$ is homotopy equivalent to the Tits building of G, which in turn has the homotopy type of a wedge of d spheres where $d = r_{\mathbb{Q}}(G) - 1$. In particular, $H_d(\partial\overline{X}) = I$, the <u>Steinberg module</u> of $G_{\mathbb{Q}}$, is free abelian of infinite rank (unless Γ is co-compact, in which case it has rank 1). It follows that when Γ is torsion free,

$$H^k(\Gamma;\mathbb{Z}\Gamma) \cong H^k_c(\overline{X}) \cong H_{n-k-1}(\partial\overline{X}) \cong \begin{cases} I, & k = n-d-1 \\ 0, & \text{otherwise.} \end{cases}$$

($n = \dim X$, H^*_c = cohomology with compact supports), so that Γ is a duality group in the sense of Bieri and Eckmann. This is equivalent to the statement that if M is any $\mathbb{Z}\Gamma$-module, then

$$H^k(\Gamma;M) \cong H_{v-k}(\Gamma;M\otimes I)$$

where $v = n-d-1$ is the cohomological dimension of Γ. (The <u>cohomological dimension</u> of Γ ($\mathrm{cd}(\Gamma)$) is the smallest integer v such that there exists a $\mathbb{Z}\Gamma$-module M with $H^v(\Gamma;M) \neq 0$.

To follow the outline above for the mapping class groups it remains only to show:

<u>Theorem 4.1</u>: The boundary ∂W is homotopy equivalent to a wedge of d-spheres where $d = 2g-2$ when $s = 0$, $g > 0$, $d = 2g-3+s$ when $s > 0$, $g > 0$ and $d = s-4$, $g = 0$.

An equivalent result is that Γ^s_g is a <u>virtual duality group</u>:

<u>Corollary 4.2</u>: If Γ is any torsion-free subgroup of finite index in Γ^s_g, then

$$H^k(\Gamma;\mathbb{Z}\Gamma) \cong \begin{cases} I = H_d(\partial W), & k = n-d-1 \\ 0, & \text{otherwise,} \end{cases}$$

where $n = \dim T^s_g = 6g-6+2s$. Thus

$$H^k(\Gamma;M) \cong H_{v-k}(\Gamma;M\otimes I)$$

with

$$v = n-d-1 = \begin{cases} 4g-5 & s = 0 \text{ and } g > 1, \\ 4g-4+s & s > 0 \text{ and } g \geq 1, \\ 1 & s = 0,\ g = 1, \\ s-3 & s > 2,\ g = 0. \end{cases}$$

The integer v is the <u>virtual cohomological dimension</u> (vcd) of Γ_g^s; that is, v is the cohomological dimension of any torsion free subgroup of finite index in Γ_g^s.

i) <u>Reduction to the case where</u> $s = 0$.

Let

$1 \to A \to B \to C \to 1$

be an exact sequence of torsion free groups. When A and C are duality groups, Bieri and Eckmann prove that B is also a duality group, $cd(B) = cd(A) + cd(C)$ and the dualizing module I_B is isomorphic to $I_A \otimes I_C$. We apply this to the exact sequence (derived from (S1) in Chapter 1):

$$1 \to \pi_1(F_g^s) \to \hat{\Gamma}_g^{s+1} \to \hat{\Gamma}_g^s \to 1 \qquad (S_2)$$

where $\hat{\Gamma}_g^s$ is torson free, finite index in Γ_g^s and $\hat{\Gamma}_g^{s+1} = \eta^{-1}(\hat{\Gamma}_g^s)$ has the same properties in Γ_g^{s+1}. This means that for fixed g if we show that Γ_g^0 is a virtual duality group then the same will hold for Γ_g^s, $s > 0$. It also means that for fixed $g > 1$ Theorem 4.1 need only be verified for $s = 0$.

We will prove Theorem 4.1 by induction on g. To start the induction we note that the cases $g = 0, 1,$ and 2 follow from more elementary results: For $g = 0$, Γ_0^3 is trivial (as are Γ_0^0, Γ_0^1 and Γ_0^2) so induction on s using (S_2) establishes Corollary 4.2 and therefore Theorem 4.1 for all Γ_0^s, $s \geq 3$. For $g = 1$, $\Gamma_1^0 \cong \Gamma_1^1 \cong SL_2\mathbb{Z}$ which is arithmetic; applications of (S_2) with $s \geq 1$ give the general case of Γ_1^s. For $g = 2$, it is known that Γ_2^0 is an extension of a finite group by Γ_0^6. Any subgroup of finite index in Γ_2^0 which is torsion free will map isomorphically onto a subgroup of Γ_0^6 having the same properties. This proves 4.2 and therefore 4.1 for Γ_2^0, thus for all Γ_2^s.

From here on we will assume $g \geq 2$. In view of the results of Chapter 3, §2 and the comments above, proving Theorem 4.1 is equivalent to showing $Z_g^0 \simeq \vee S^{2g-2}$.

ii) Z_g^0 <u>is</u> $2g-3$ <u>connected</u>.

The subcomplex $A_\infty \subset A$ was defined to be the simplices of A corresponding to arc-systems which do not fill the surface F. It takes $2g$ curves to fill F so A_∞ contains the $2g-2$ skeleton of A. It can be shown that $A \simeq W$ so A is contractible and A_∞ is $2g-3$ connected.

By forgetting the base point we may define a map $\eta : Z_g^1 \to Z_g^0$. This map has a right inverse $\omega : Z_g^0 \to Z_g^1$ defined as follows. Choose a hyperbolic metric on F so that all simple closed curves in F are represented by geodesics. Select a point $p \in F$ such that no simple closed geodesic passes through p; this gives a map from the curve systems in F to the curve systems in $F-p$ and defines ω. Clearly $\eta \circ \omega = 1$, so η_* is surjective on each π_k.

By the remarks above the following will imply Z_g^1 and Z_g^0 are $2g-3$ connected.

Lemma 4.3: The complexes A_∞ and Z_g^1 are (equivalently) homotopy equivalent.

Proof: If $C \subset F_g - \{p\}$ is any simple closed curve, define $A_C = \{\langle\alpha_0,\dots,\alpha_k\rangle : \alpha_i \cap C'$ is empty for some C' isotopic to $C\}$. Clearly $A_C \subset A_\infty$ and $\bigcup_C A_C = A_\infty$. A_C may be identified with the arc-system complex of the component of $F-C$ which contains p; it is therefore contractible. Finally, for any $C_1,\dots,C_k$ (disjoint or not), $\bigcap_{i=1}^{k} A_{C_i}$ is either empty or contractible. These facts mean that $A_\infty \simeq N$, the nerve of the cover $\{A_C\}$.

Let $C_0,\dots,C_k$ be a collection of $k+1$ simple closed curves (not necessarily disjoint, but nontrivial and nonisotopic) in $F-p$. Isotope the curves to have minimal geometric intersection and let $F(C_0,\dots,C_k) = F -$ (open regular neighborhood of $\cup C_i$). Set $F_0(C_0,\dots,C_k)$ equal to the component of $F(C_0,\dots,C_k)$ that contains p. The nerve N may be described combinatorially as the complex which has a k-cell $[C_0,\dots,C_k]$ for each isotopy class of simple closed curves $C_0,\dots,C_k$ such that $F_0(C_0,\dots,C_k)$ is not simply connected and has the usual face relations given by inclusion. From this description it is clear that Z_g^1 may be identified with a subcomplex of N

Let N^o and Z^o denote the first barycentric subdivisions of N and Z_g^1 respectively; we define a retraction $r : N^o \to Z^o$ as follows. Given a collection $C_0,\dots,C_k$, set $\partial' F(C_0,\dots,C_k)$ equal to the curve-system in F obtained from $\partial F(C_0,\dots,C_k)$ by omitting null-homotopic curves and redundancies. Let $B(C_0,\dots,C_k)$ denote the barycenter of the cell $[C_0,\dots,C_k]$ of N and, when $C_0,\dots,C_k$

form a curve system, the barycenter of the cell $<C_0,\dots,C_k>$ of Z_g^1. Now set $r(B(C_0,\dots,C_k)) = B(\partial' F(C_0,\dots,C_k))$ and extend linearly. Clearly $r|Z^o$ is the identity. To see that $r \simeq 1$ triangulate $N^o \times I$ by identifying $N^o \times 0$ and $N^o \times 1$ with N^o and joining each

$$[C_0] < [C_1] < \dots < [C_k]$$

in $N^o \times 1$ to each chain

$$[C_0'] < \dots < [C_\ell']$$

in $N^o \times 0$ whenever $C_\ell' \subset C_0$. Since $\partial' F(C_0,\dots,C_k)$ is disjoint from each C_i it is clear that $1|N^o \times 0 \coprod r|N^o \times 1$ extends linearly to $N^o \times I$, providing the needed homotopy.

iii) Z_g^0 <u>has the homotopy type of a</u> 2g-2 <u>complex</u>

To finish the proof of Theorem 4.1 we must now show that Z_g^0, which has dimension $3g-4$, has the homotopy type of a complex of dimension $2g-2$ (recall we are assuming $g \geq 2$). Let Z^o denote the first barycentric subdivision of Z_g^0, we build Z^o piece by piece, including the subcomplexes spanned by vertices of descending weight.

Let X_k be the subcomplex of Z^o consisting of simplices whose vertices have weight $\geq k$. Assume Theorem 4.1 for all Z_h^s with $h < g$ and assume that X_{k+1} has been shown to be homotopy equivalent to a complex of dimension $\leq 2g-2$. X_k is obtained from X_{k+1} by adding the simplices which have exactly one vertex of weight k (since no two vertices of a simplex of Z^o have the same weight). Let v be a vertex of $X_k - X_{k+1}$ corresponding to the curves $C_0,\dots,C_k$. If $F^1,\dots,F^t$ are the components of the surface obtained by splitting F along $\{C_i\}$ and $Z(F^i)$ denotes the complex of curve-systems on F^i, then the link of v in X_{k+1} is easily identified with the join of the $Z(F^i)$. If F^i has genus g_i and r_i boundary components, then $g = \sum g_i + k - t + 2$ with $g_i < g$ for all i and $\sum r_i = 2k+2$. Since each $Z(F^i)$ is by assumption homotopy equivalent to a complex of dimension $\leq 2g_i + r_i - 3$, the link of v is homotopy equivalent to one of dimension $\leq \sum(2g_i + r_i - 3) + (t-1) = 2g-3$. This verifies that X_k is homotopically of dimension $\leq 2g-2$ for each k; since $X_0 = Z^o$ the proof of Theorem 4.1 is now complete.

§2. The Mapping Class Group is not Arithmetic

Suppose that the group $\Gamma = \Gamma_g^s$ is an arithmetic subgroup of the linear algebraic group G. Then Corollary 4.2 can be combined with the results of Borel and Serre to see that if n is the dimension of the symmetric space X associated to G, then $n - r_{\mathbb{Q}}(G) = \mathrm{vcd}(\Gamma)$. The number $r_{\mathbb{Q}}(G)$ can be computed directly from Γ; it is the maximal rank of an abelian subgroup of Γ. In [BLM] this is shown to be $3g-3+s$, so $n = 7g-8$ when $g > 1, s = 0$ and $n = 7g-7+2s$ for $g > 0$. Thus Teichmüller space cannot be identified with X when $s \geq 0$, $g > 2$ or $s > 0, g = 2$.

This still does not mean Γ cannot be arithmetic; to prove it cannot be we argue as follows (this argument was shown to us by Bill Goldman). Suppose first that G has rank 1 so that X is hyperbolic space. Two elements of infinite order in the group of isometries of X then commute if and only if they have the same fixed point set on the sphere at ∞. This means that commuting is an equivalence relation on the elements of infinite order in Γ. For the mapping class group this is absurd: simply take curves C_0, C_1 and C_2 with C_0 disjoint from C_1 and C_2 but C_1 intersecting C_2 (say in one point). If τ_i denotes the Dehn twist on C_i, then τ_0 commutes with τ_1 and τ_2 but τ_1 and τ_2 do not commute.

An alternative argument for the rank 1 case goes as follows. If Γ is cocompact, the dualizing module I will be isomorphic to $\mathbb{Z}$. It is not hard to show this is not true; on the contrary, it has infinite rank. If Γ is not cocompact it must at least be cofinite volume. The V-manifold X/Γ will have cusps, modeled on horoballs/Γ. The boundary horosphere Σ of one of these horoballs has a Euclidean structure and since Σ/Γ is compact it is covered by an $n-1$ dimensional torus. This implies Γ has an abelian subgroup of rank $n-1$ which is impossible by the dimension count above.

Next consider the case where G has rank ≥ 2. As mentioned earlier, property (8) and the existence of the Torelli group say that G cannot be simple. Suppose then that G is semisimple; a stronger version of (8), also proven by Margulis, says that if Γ is an irreducible lattice in G, then once again any normal subgroup in G is either finite (and central) or finite index. This means Γ must be reducible, so there is a subgroup of finite index in Γ which is a direct product of infinite groups. An analysis of the centralizers of the elements of Γ (compare [McCar]) shows this is not possible. Finally, if G is not semisimple we need to look at solvable subgroups of Γ. A theorem of Birman, Lubotsky and McCarthy [BLM] says that every solvable subgroup of Γ is virtually abelian. Such a subgroup will not be normal in Γ, so the map $G \to G/\mathrm{rad}(G)$ imbeds Γ in the semisimple group $G/\mathrm{rad}(G)$; this is impossible.

Chapter 5. The Weil-Petersson Geometry of Teichmüller Space

In this chapter we will describe some results of Scott Wolpert on the geometry of Teichmüller space ([Wol 1]-[Wol 5]). The theme we will be following is the counterpart to that of chapter 4, namely: How close is Teichmüller space to being a symmetric space? We will translate this into the question: How much of the formal geometry of a symmetric space does Teichmüller space have? The metric we will study on T_g^s is the Weil-Petersson metric; it is Kähler and we will see that its Hermitian and symplectic geometry arise from the hyperbolic geometry of the surface. The metric is also invariant under the action of the mapping class group Γ_g^s; on M_g^s is it not complete, rather it admits a continuous extension to the Deligne-Mumford compactification $\overline{M}_g^s$ ([Mas]). The corresponding Kähler form ω_{wp} extends to $\overline{\omega}_{wp}$ on $\overline{M}_g^s$; in §3 we will show how Wolpert uses $\overline{\omega}_{wp}$ to give an analytic proof that $\overline{M}_g^s$ is projective.

§1. The Symplectic Geometry of the Weil-Petersson Form

We begin with the definition of the Weil-Petersson metric on T_g^s. Let R be a Riemann surface and let λ be the hyperbolic line element on R. Teichmüller space is a complex manifold and the holomorphic cotangent space at R may be identified with $Q(R)$, the space of integrable holomorphic quadratic differentials on R (tensors of type $dz \otimes dz$). If $\phi, \psi \in Q(R)$, the Hermitian product

$$\langle \phi, \psi \rangle = \frac{1}{2} \int_R \phi \, \bar{\psi} \, \lambda^{-2}$$

defines the Weil-Petersson metric at R. This metric is Kähler; its corresponding Kähler form is denoted ω_{wp}. The first thing we will do is to give Wolpert's formula for ω_{wp} in terms of Fenchel-Nielsen coordinates.

i) ω_{wp} in Fenchel-Nielsen Coordinates.

Let $C = \{C_1, \ldots, C_n\}$ be a maximal curve system in F and let (τ_i, ℓ_i) be the corresponding Fenchel-Nielsen coordinates for T_g^s.

Theorem 5.1 $\omega_{wp} = \sum_i d\ell_i \wedge d\tau_i$.

Several things about this statement are surprising. First of all, the Weil-Petersson metric is Kähler, while Fenchel-Nielsen coordinates are only real analytic; the simplicity of the formula is therefore unexpected. Secondly, the Weil-Petersson metric is invariant under the action of the mapping class group, so $\sum d\ell_i \wedge d\tau_i$ must be also. Actually, Theorem 5.1 says more since it shows that $\sum d\ell_i \wedge d\tau_i$ is independent of the curve-system C. By contrast the change of coordinates from one curve-system to another can be quite complicated.

Theorem 5.1 is a consequence of the duality formula which was used already in chapter 3. To state this we must first define the Fenchel-Nielsen <u>twist vector fields</u> t_C. Let $(X,[f])$ represent a point of Teichmüller space with X hyperbolic and let α be the closed geodesic on X representing the free homotopy class $f^{-1}(C)$ where $C \subset F$ is a nontrivial simple closed curve. Cut X along α, rotate <u>one side</u> of the cut and then reglue the sides. The hyperbolic structure in the complement of the cut extends naturally to a hyperbolic structure on the new surface. Varying the amount of rotation gives a flow on T_g^s and the twist vector field t_C (sometimes denoted t_α) is the tangent vector field to this flow. We will always normalize t_C so that the hyperbolic displacement of two points on opposite sides of the geodesic α increases at unit speed (thus for example a full rotation about α occurs at time $t =$ length of α).

Let $C \subset F$ be any nontrivial simple closed curve. The <u>Duality Formula</u> now states:

<u>Theorem 5.2</u>: $\omega_{wp}(t_C, \ ) = -d\ell_C$.

ii) <u>Proof of Theorem 5.2</u>

To prove this formula we introduce $H(R)$, the space of <u>harmonic Beltrami differentials</u> on R. An element $\mu \in H(R)$ is a tensor of type $\frac{\partial}{\partial z} \otimes d\bar{z}$ and is harmonic with respect to the Laplace-Beltrami operator for the hyperbolic metric on R. The holomorphic tangent space at R may be identified with $H(R)$ and the Weil-Petersson metric on T_g^s has the dual expression

$$\langle \mu, \nu \rangle = \int_R \mu\bar{\nu}\, \lambda^2,$$

$\mu, \nu \in H(R)$ and λ the hyperbolic line element as before.

The underlying Riemannian structure to $\langle \ , \ \rangle$ is of course given by the symmetric tensor

$$g(\mu,\nu) = 2\mathrm{Re}\langle \mu,\nu \rangle,$$

and the Weil-Petersson Kähler form is defined by the equation

$$\omega_{wp}(\mu,\nu) = g(J\mu,\nu)$$

where J is the complex structure on $\mathcal{T}_g^s$. This means that $\omega_{wp}(t_C, \ )$ is the Riemannian dual of Jt_C.

The next step is to write down formulas for t_C and $d\ell_C$ in terms of Poincaré series. Let $X = \mathbb{H}^2/\Gamma$ where Γ is Fuchsian and let α be the simple closed geodesic representing C in X. When $A = \begin{pmatrix} a & b \\ c & d \end{pmatrix}$ represents α in $\Gamma < PSL_2\mathbb{R}$, define $\Omega_A(\zeta) = (tr^2A-4)(c\zeta^2+(d-a)\zeta-b)^{-2}$. If $\langle A \rangle$ denotes the infinite cyclic group generated by A, we set

$$\Theta_C = \sum_{B \in \Gamma/\langle A \rangle} \Omega_{B^{-1}AB};$$

this is a relative Poincaré series and it converges uniformly and absolutely on compact sets. A formula of Gardiner [G] expresses $d\ell_C$ in terms of Θ_C: if $\mu \in H(X)$ represents a tangent vector, then

$$\mathrm{Re}(d\ell_C(\mu)) = \frac{2}{\pi} \mathrm{Re} \int_X \mu\,\Theta_C.$$

In particular, we may write $-d\ell_C = -\frac{2}{\pi}\Theta_C$. On the other hand, Wolpert uses the Bers embedding of $\mathcal{T}_g^s$ into the vector space of Γ invariant holomorphic quadratic differentials to show that

$$t_C = \frac{i}{\pi}(\mathrm{Im} z)^2\,\overline{\Theta}_C$$

[Wol 1]. Therefore $Jt_C = -\frac{1}{\pi}(\mathrm{Im} z)^2\,\overline{\Theta}_C$ so $-d\ell_C$ is also the Riemannian dual to Jt_C. Theorem 5.2 follows. □

iii) <u>Proof of Theorem 5.1</u>

Next we derive Theorem 5.1 from 5.2. First note that if $\{C_i\}$ is a partion of F giving Fenchel-Nielsen coordinates $\{\tau_i, \ell_i\}$, then the twist vector fields t_{C_i} are just the coordinate vector fields $\frac{\partial}{\partial\tau_i}$. Furthermore, the duality formula implies that ω_{wp} is invariant under any twist flow. Combining this with the fact that the coordinate vector fields $\{\frac{\partial}{\partial\tau_i}, \frac{\partial}{\partial\ell_i}\}$ commute, it follows that the coefficients of ω_{wp} in the basis $\{d\tau_i\wedge d\tau_j, d\ell_i\wedge d\tau_j, d\ell_i\wedge d\ell_j\}$ are independent of τ_i. From this one can show that it suffices to compute ω_{wp} at those surfaces X which admit an orientation reversing isometry ρ fixing the partition $\{C_i\}$ $(\rho(\alpha_i) = \alpha_i$ for each α_i where $\{\alpha_i\}$ are the geodesics representing $\{C_i\}$ on $X)$. The functions ℓ_{C_i} are invariant under ρ since the length of a curve does not depend on the orientation of the surface. On the other hand, the twist parameter

τ_i does depend on this orientation since right and left are reversed by ρ. One makes this precise by showing:

$$\rho^* d\ell_{C_i} = d\ell_{C_i} \quad \text{and} \quad \rho^* d\tau_{C_j} = -d\tau_{C_j} + \frac{n_j}{2} d\ell_{C_j}$$

for some integers n_j. Since ρ corresponds to an element of the mapping class group which acts antiholomorphically,

$$\rho^*\omega_{wp} = -\omega_{wp}.$$

Now since the coefficients of $d\tau_i \wedge d\tau_j$ and $d\ell_i \wedge d\ell_j$ are even relative to a ρ substitution, while ω_{wp} is odd, these coefficients are identically zero. Finally, the fact that the coefficient of $d\ell_i \wedge d\tau_j$ is the Kronecker delta δ_{ij} follows directly from 5.2. □

iv) Consequences of Theorems 5.1 and 5.2

The first consequence of Theorem 5.2 is that the vector fields t_C are Hamiltonian for ω_{wp}; that is, the Lie derivative $L_{t_C}\omega_{wp}$ vanishes. This follows from the general formula $L_X\omega_{wp} = (d\omega_{wp})(X, \ , \ ) + d(\omega_{wp}(X \ , \ ))$, the fact that ω_{wp} is Kähler (thus $d\omega_{wp} = 0$) and the duality formula. Thus we see that ω_{wp} and the vector fields t_C define a symplectic geometry on $\mathcal{M}_g^s$ and $\mathcal{T}_g^s$ (later we will see that ω_{wp} extends smoothly to $\overline{\mathcal{M}}_g^s$ where it remains symplectic). By analogy with symmetric spaces we may use ω_{wp} to define a Lie algebra: just take the vector space of all vector fields X on $\mathcal{T}_g^s$ such that $L_X\omega_{wp} = 0$. It can be shown that this Lie algebra is generated over the C^∞ functions by the L_{t_C}; however, it is infinite dimensional.

An idea suggested by the preceeding Theorems is that the hyperbolic geometry on the surface is reflected in the symplectic geometry of Teichmüller space. There are three main formulas that come out; they are the cosine formula, the sine-length formula and the Lie bracket formula. We state them without proof.

Cosine Formula

Let C_1 and C_2 be two nontrivial simple closed curves in F; then at $X \in \mathcal{T}_g^s$

$$t_{C_1}\ell_{C_2} = \omega_{wp}\left(t_{C_1}, t_{C_2}\right) = \sum_{p \in \alpha \# \beta} \cos\Theta_p$$

where α and β are the geodesics in X representing C_1 and C_2, $\alpha \# \beta = \alpha \cap \beta$ unless $\alpha = \beta$ in which case it is empty, and Θ_p denotes the angle between α and β. Here $t_{C_1}\ell_{C_2}$ means the Lie derivative of ℓ_{C_2} by the vector field t_{C_1}.

Sine-Length Formula

Let C_0, C_1, C_2 be three nontrivial simple closed curves in F and let α, β, γ represent C_0, C_1, C_2 respectively in $X \in \mathcal{T}_g^s$. Then

$$t_{C_0} t_{C_1} \ell_{C_2} = \sum_{(p,q) \in \alpha\#\gamma \times \beta\#\gamma} \frac{e^{m_1} + e^{m_2}}{2(e^{\ell_\gamma} - 1)} \sin\Theta_p \sin\Theta_q$$

$$- \sum_{(r,s) \in \alpha\#\beta \times \beta\#\gamma} \frac{e^{n_1} + e^{n_2}}{2(e^{\ell_\beta} - 1)} \sin\Theta_r \sin\Theta_s$$

where the two possible routes from p to q along γ have length m_1 and m_2 and the two routes from r to s along β have length n_1 and n_2.

Lie Bracket Formula

Renormalize the vector fields t_C by setting $T_C = 4(\sinh \frac{\ell_C}{2}) t_C$. Then, with notation as above,

$$[T_{C_1}, T_{C_2}] = \sum_{p \in \alpha\#\beta} T_{\alpha_p\beta^+} - T_{\alpha_p\beta^-}$$

where $\alpha_p\beta^+$ and $\alpha_p\beta^-$ are the curves in F corresponding to the configurations in Figure 5.1.

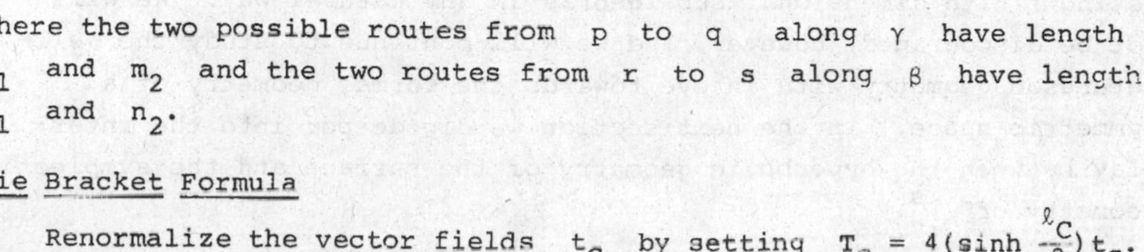

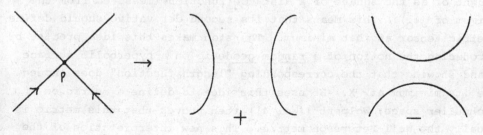

Figure 5.1

Actually, in the third formula we have done something illegal in that the curves $\alpha_p\beta^+$ and $\alpha_p\beta^-$ may not be simple. Nevertheless, the formula makes sense because it is possible to define the functions ℓ_C and the vector fields t_C even when C is not simple. (It is clear how to do this for ℓ_C, and t_C is given by duality). This suggests that a Lie algebra over the integers might be constructed as follows. Let $\hat{\pi}$ denote the conjugacy classes in the fundamental group of the surface F_g and let $\mathbb{Z}\hat{\pi}$ denote the free abelian group on the elements of $\hat{\pi}$. Fix a hyperbolic metric on F; each element of $\hat{\pi}$ is represented by a unique closed (but not necessarily simple) geodesic. Goldman [Gold] defines a bracket on $\mathbb{Z}\pi$ by setting

$$[\alpha,\beta] = \sum_{p\in\alpha\#\beta} \alpha_p\beta^+ - \alpha_p\beta^-$$

and extending linearly. The proof that [,] satisfies the hypotheses to be a Lie bracket on $\mathbb{Z}\hat{\pi}$ is purely topological.

Unfortunately, all these definitions give infinite dimensional Lie algebras and there is no indication that they contain any interesting finite dimensional subalgebras in any natural way. We will not be discouraged, however, and we will continue to study the Weil-Petersson geometry with an eye towards the formal geometry of a symmetric space. In the next section we dig deeper into the interplay between the hyperbolic geometry of the surface and the symplectic geometry of T_g^s.

§2. The Thurston-Wolpert Random Geodesic Interpretation of the Weil-Petersson Metric

The key step in Kerckhoff's proof of the Nielsen conjecture is his observation that the geodesic length functions are convex along earthquake paths. While it would be out of place here to discuss either the Nielsen conjecture or earthquakes, we can still abstract from this the idea that a geodesic length function ℓ_C can be thought of as the square of a distance function (measured from the minimum of ℓ_C). This means that its second derivative should define a metric tensor at that minimum. Thurston makes this idea precise by introducing the notion of a <u>random geodesic</u> on a hyperbolic surface X and showing that the corresponding "length function" does indeed have its minimum at X. He uses this idea to define a metric on Teichmüller space; Wolpert ([Wol 4]) then proves that this metric is actually the Weil-Petersson metric. This new interpretation of the metric proves to be quite valuable for it gives the following remarkable formula for the complex structure on T_g:

$$Jt_C = 3\pi(g-1)\lim_j \frac{[t_C, t_{\beta_j}]}{\ell_{\beta_j}}$$

where β_j are random geodesics.

i) Definition of the Random Geodesic

Let X be a hyperbolic surface and $\beta_j \subset X$ a sequence of closed geodesics (not necessarily simple). If T_1X denotes the unit tangents to X, then each β_j has a unique lift $\tilde{\beta}_j$ to T_1X. Define $\{\beta_j\}$ to be uniformly distributed if for all open sets $U \subset T_1X$

$$\lim_j \frac{\ell(\tilde{\beta}_j \cap U)}{\ell(\tilde{\beta}_j)} = \frac{\text{Volume }(U)}{\text{Volume}(T_1X)}.$$

Here we identify T_1X with $T_1\mathbb{H}^2/\Gamma$ where $X = \mathbb{H}^2/\Gamma$ and compute volume via the isomorphism $T_1\mathbb{H}^2 \cong PSL_2\mathbb{R}$. Let $\langle\ ,\ \rangle$ denote the Weil-Petersson metric.

Theorem 5.3: For any nontrivial simple closed curves C_1, C_2,

$$\langle t_{C_1}, t_{C_2}\rangle = 3\pi(g-1) \lim_j \frac{t_{C_1} t_{C_2} \ell_{\beta_j}}{\ell_{\beta_j}},$$

where $\{\beta_j\}$ is uniformly distributed on X.

In the formula we are writing ℓ_{β_j} when we really mean $\ell_{f(\beta_j)}$ where $f\colon X \to F$ is the marking on X.

ii) Outline of the proof of Theorem 5.3

First we explain why $\langle t_{C_1}, t_{C_2}\rangle_T = \lim_j \frac{1}{\ell_{\beta_j}} t_{C_1} t_{C_2} \ell_{\beta_j}$ is symmetric. Since $[t_{C_1}, t_{C_2}]$ is a tangent vector for every C_1 and C_2, this will follow if we show $\lim_j \frac{1}{\ell_{\beta_j}} V \ell_{\beta_j} = 0$ for any tangent vector V. Since twist vector fields span the tangent space to T_g^s, it suffices to show $\lim_j \frac{1}{\ell_{\beta_j}} t_C \ell_{\beta_j} = 0$ for every C. But the β_j are uniformly distributed in T_1X, so for any closed geodesic $\alpha \subset X$, in the limit each intersection of α with β_j of angle Θ is accompanied by another of angle $\pi - \Theta$. Applying the cosine formula completes the argument.

The backbone of the rest of the argument is the fact that $\langle\ ,\ \rangle_T$ is constructed naturally with respect to the $PSL_2\mathbb{R}$ geometry on the unit tangent bundle T_1X. To make use of this, Wolpert observes that there exist tensors K_1 and K_2 on X such that for any two holomorphic quadratic differentials ϕ, ψ on X

even continuous in complex coordinates on $\overline{M}_g$. In fact $\overline{\omega}_{wp}$ is only a current and thus its positivity and rationality are difficult to check.

The approach consists of three parts; the extension of the Kähler form to $\overline{M}_g$, the rationality of the extension and of course the positivity of the resulting line bundle.

i) The extension of ω_{wp} to $\overline{M}_g$

Masur [Mas] was the first to consider the extension of ω_{wp} to $\overline{M}_g$. At a generic point of $D = \overline{M}_g - M_g$ with normal coordinate z, he gave the formula

$$\overline{\omega}_{wp} \sim \frac{i/2\, dz \wedge d\overline{z}}{|z|^2 (\log 1/|z|)^3},$$

showing that the extension is not continuous in the complex coordinates. By contrast, since Fenchel-Nielsen coordinates on $\overline{M}_g$ are obtained simply by allowing the ℓ_i to be 0, the formula $\omega_{wp} = \sum d\ell_i \wedge d\tau_i$ shows that ω_{wp} extends smoothly in Fenchel-Nielsen coordinates. This means that the complex structure $\overline{M}_g^{\mathbb{C}}$ and the real-analytic structure $\overline{M}_g^{FN}$ are not subordinate to a common smooth structure on $\overline{M}_g$. Nevertheless, in [Wol 5] it is proved that the identity map $\overline{M}_g^{FN} \to \overline{M}_g^{\mathbb{C}}$ is Lipschitz continuous. From the point of view of cohomology this means we can integrate $\overline{\omega}_{wp}$ over 2-cycles in Fenchel-Nielsen coordinates and use the answer to study $\overline{M}_g^{\mathbb{C}}$.

Masur's formula shows that the form $\overline{\omega}_{wp}$ has singularities along D. Wolpert deals with this by first showing that $\overline{\omega}_{wp}$ is a closed, positive (1,1) current and then that $\overline{\omega_{wp}}$ is the limit of smooth, closed, positive (1,1) forms.

ii) Rationality of $\frac{1}{\pi^2}\overline{\omega}_{wp}$

The second part of the proof is to show that $\frac{1}{\pi^2}\overline{\omega}_{wp}$ is rational. By Theorem 7.2 below ([Hl]), $H_2(M_g;\mathbb{Q})$ is rank 1 for $g \geq 3$. An application of Mayer-Victoris then shows $H_2(\overline{M}_g;\mathbb{Q}) \cong H^{6g-8}(\overline{M}_g;\mathbb{Q})$ has rank $2 + [g/2]$. We will describe dual bases for H_2 and H_{6g-8}.

The divisor $D = \overline{M}_g - M_g$ is the sum of $1 + [g/2]$ components $D_0, \ldots, D_{[g/2]}$ where D_i generically consists of the surfaces with a node which are obtained by collapsing a simple closed curve to a point. For D_0 the curve is nonseparating while for D_i, $i > 0$, the curve separates the surface into pieces of genus i and $g-i$. The Poincaré dual of $\overline{\omega}_{wp}$ and $D_0, \ldots, D_{[g/2]}$ give $2+[g/2]$ classes in $H_{6g-8}(\overline{M}_g)$.

$$\langle\phi,\psi\rangle_T = \int_X \phi\bar{\psi}K_1 + \phi\psi K_2.$$

K_1 must be a (-1,-1) tensor and K_2 a (-3,1) tensor and by the naturality of $\langle\ ,\ \rangle_T$ both must be $PSL_2\mathbb{R}$ invariant. The only possibility is that K_1 is a multiple of the hyperbolic area element and K_2 is zero. Thus $\langle\ ,\ \rangle_T$ is a multiple of $\langle\ ,\ \rangle_{wp}$.

iii) Description of the Complex Structure on T_g^s

Let C be any nontrivial simple closed curve and $\{\beta_j\}$ a sequence of uniformly distributed geodesics in X. Then Wolpert uses Theorem 5.3 to show:

Theorem 5.4: $Jt_C = 3\pi(g-1)\lim_j \frac{1}{\ell_{\beta_j}}[t_C, t_{\beta_j}]$ where J is the complex structure on T_g^s.

The proof of this uses not only 5.3 but also the Lie Bracket formula, the duality formula and the formula

$$t_{\alpha_1}t_{\alpha_2}\ell_\beta + t_{\alpha_2}t_\beta\ell_{\alpha_1} + t_\beta t_{\alpha_1}\ell_{\alpha_2} = 0$$

from [Wol 3]. By skew symmetry $t_{\alpha_1}\ell_{\alpha_2} = -t_{\alpha_2}\ell_{\alpha_1}$, so $t_{\alpha_1}t_{\alpha_2}\ell_\beta + [t_{\alpha_2},t_\beta]\ell_{\alpha_1} = 0$. Dividing by ℓ_β and taking β in a sequence which is uniformly distributed gives

$$\langle t_{\alpha_1}, t_{\alpha_2}\rangle + 3\pi(g-1)\omega_{wp}(\lim_j \frac{1}{\ell_{\beta_j}}[t_{\alpha_2}, t_{\beta_j}], t_{\alpha_1}) = 0.$$

Since α_1 is arbitrary $t_\alpha + 3\pi(g-1)\, J \lim_j \frac{1}{\ell_{\beta_j}}[t_\alpha, t_{\beta_j}] = 0.$

Theorem 5.4 now follows. □

§3. The Projective Embedding of $\overline{M}_g$

As a final illustration of the importance of the Weil-Petersson geometry we will give a sketch of Wolpert's beautiful proof that $\overline{M}_g$ is projective. The outline is very simple: first he shows that ω_{wp} extends to $\overline{\omega}_{wp}$ on $\overline{M}_g$. Next he proves that $\frac{1}{\pi^2}\overline{\omega}_{wp}$ is rational and therefore $\frac{n}{\pi^2}\overline{\omega}_{wp}$ is c_1 of a line bundle L for some positive integer n. Since ω_{wp} is Kähler, L is postive; the Kodaira theorem now provides the imbedding $\overline{M}_g \to \mathbb{C}P^n$.

Of course things are not really as simple as all that. The extension $\overline{\omega}_{wp}$ is smooth in Fenchel-Nielsen coordinates but is not

For $H_2(\overline{M}_g)$ take one of the $2+[g/2]$ configurations of Figure 5.2 and consider the subset of D described by allowing the structure

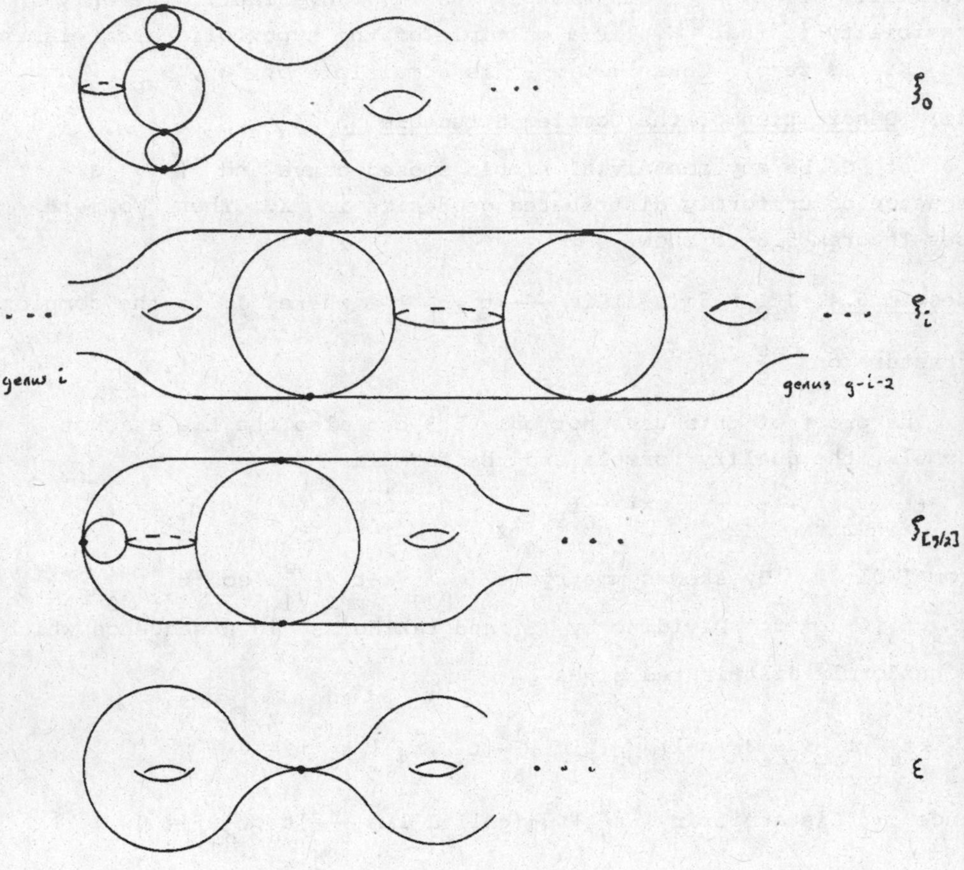

Figure 5.2

on the component labeled A to vary while that on the rest of the surface remains fixed. Each of these describes an analytic 2-cycle in D; ε is isomorphic to $\overline{M}_1^1$ while $\xi_0,\ldots,\xi_{[g/2]}$ are isomorphic to $\overline{M}_0^4$. The next thing to do is to compute the intersection matrix between these two collections; it will be nonsingular, showing that both sets are actually bases. This is straightforward except for evaluating the integrals $\int_\varepsilon \overline{\omega}_{wp}$ and $\int_{\xi_i} \overline{\omega}_{wp}$. Now Wolpert computes directly that $\int_{\overline{M}_1^1} \overline{\omega}_{wp} = \pi^2/6$. Passing to a 4-fold cover gives

$\int_{\overline{M}_1^4} \overline{\omega}_{wp} = 2\pi^2/3$ and since F_1^4 double covers F_0^4, $\int_{\overline{M}_0^4} \overline{\omega}_{wp} = \pi^2/3$.

Using Theorem 5.1 it is easy to see that $\overline{\omega}_{wp}$ restricts, i.e. $\overline{\omega}_{wp}|\varepsilon = \overline{\omega}_{wp}$ on $\overline{M}_1^{\,1}$ and $\overline{\omega}_{wp}|\xi_i = \overline{\omega}_{wp}$ on $\overline{M}_0^{\,4}$. This shows not only that the collections are bases, but simultaneously evaluates $\overline{\omega}_{wp}$ and shows $\frac{1}{\pi^2}\overline{\omega}_{wp}$ is rational.

iii) __The Positive Line Bundle__

Now $\frac{1}{\pi^2}\overline{\omega}_{wp}$ is a rational, closed positive (1,1) current on $\overline{M}_g$. We would like to say that $\frac{n}{\pi^2}\overline{\omega}_{wp}$ is c_1 of a positive line bundle L on $\overline{M}_g$. Recall that for any L, $c_1(L) = \frac{1}{2\pi i}\partial\overline{\partial}\log\|s\|^2$ where s is any section of L and $\|\ \|$ is a metric for L. Heuristically, if $\|s\|^2 = e^{1/\log(1/|z|^2)}$, $|z|< 1$, then we find

$$\partial\overline{\partial}\log\|s\|^2 = \partial\overline{\partial}\ \frac{1}{\log(1/|z|^2)} = \frac{2}{|z|^2(\log(1/|z|^2))^3}\ ;$$

the principal term of Masur's formula. Now by our calculation $e^{1/\log(1/|z|^2)}$ is a C^0 metric of positive curvature in the sense of currents. Similarly Wolpert finds that $\overline{\omega}_{wp}$ is the positive curvature form of a C^0 metric for a line bundle over $\overline{M}_g$. An argument of Richberg is then used to replace the metric in the line bundle by a smooth metric with positive corvature form. This completes the argument.

Chapter 6 Stability of the Homology of Γ_g^s

We return now to a direct study of $H_*(\Gamma_g^s)$. A natural question to ask about Γ_g^s is whether its homology stabilizes as $g \to \infty$. This kind of result is known (rationally, at least) for many classes of arithmetic groups, e.g. SL_n and Sp_{2g} ([B]), and the techniques to prove it are readily available ([C],[Q],[V],[W]). We will now combine these techniques with the results of Chapter 2 and 4 to show that in fact $H_k(\Gamma_g^s)$ is independent of g when $g >> k$.

Let $F_{g,r}^s$ denote an oriented surface of genus g with r boundary components and s punctures. The mapping class group $\Gamma_{g,r}^s$ is defined to be the group of all isotopy classes of homeomorphisms of F which are the identity on ∂F and fix the punctures individually. We emphasize that isotopies must fix ∂F pointwise, otherwise there would be no distinction between $\Gamma_{g,r}^s$ and $\Gamma_{g,0}^{r+s}$. Define three maps

$$\phi: F_{g,r}^s \to F_{g,r+1}^s \quad , r \geq 1,$$

$$\psi: F_{g,r}^s \to F_{g+1,r-1}^s \quad , r \geq 2,$$

$$\eta: F_{g,r}^s \to F_{g+1,r-2}^s \quad , r \geq 2$$

as follows. For ψ sew a pair of pants (a copy of the surface $F_{0,3}^0$) to $F_{g,r}^s$ along two components of its boundary. For ϕ sew a pair of pants to $F_{g,r}^s$ along one component of its boundary. Finally, for η sew two components of $\partial F_{g,r}^s$ together. The maps ϕ, ψ, η induce maps of mapping class groups (extend by the identity on $F_{0,3}^0$ for ϕ and ψ; η is obvious); let ϕ_*, ψ_*, η_* be (respectively) the maps they induce on homology. We may now state the main result of $[H_2]$:

<u>Theorem</u> <u>6.1</u>: $\phi_*: H_k(\Gamma_{g,r}^s) \to H_k(\Gamma_{g,r+1}^s)$ is an isomorphism for $g \geq 3k-2$

$\psi_*: H_k(\Gamma_{g,r}^s) \to H_k(\Gamma_{g+1,r-1}^s)$ is an isomorphism for $g \geq 3k-1$

$\eta_*: H_k(\Gamma_{g,r}^s) \to H_k(\Gamma_{g+1,r-2}^s)$ is an isomorphism for $g \geq 3k$.

By combining these maps in various ways we can see that Theorem 6.1 implies $H_k(\Gamma_{g,r}^s)$ is independent of g and r as long as $g \geq 3k+1$.

For the moduli spaces this says that $H_k(M_g^s; \mathbb{Q})$ does not depend on g when $g \geq 3k+1$.

Before giving the proof of 6.1 we make some observations due to Ed Miller [M_i]. Let $\Lambda_{g,1} = \mathrm{Diff}^+(F_{g,1})$ be the group of diffeomorphisms fixing the boundary component of F pointwise. Taking boundary connected sum defines a natural homomorphism

$$\Lambda_{g,1} \times \Lambda_{h,1} \to \Lambda_{g+h,1}$$

which induces a product on classifying spaces

$$\lambda\colon B\Lambda_{g,1} \times B\Lambda_{h,1} \to B\Lambda_{g+h,1}.$$

Now set $A = \varinjlim H_*(B\Lambda_{g,1}; \mathbb{Q})$ where the limit is defined using the natural inclusions $\to \Lambda_{g,1} \to \Lambda_{g+1,1} \to \cdots$. Theorem 6.1 implies A is finite type; therefore A, under the product λ_* induced from λ, is a commutative, cocommutative Hopf algebra of finite type. A theorem of Milnor and Moore [MM] then implies:

<u>Corollary</u> <u>6.2</u>: A is the tensor product of a polynomial algebra on even dimensional generators with an exterior algebra on odd dimensional generators.

We move now to the proof of Theorem 6.1.

§1) The Cell Complexes

The first half of the proof is an excursion through a maze of different cell complexes which are constructed from configurations of arcs and circles in a surface. There are six of these; their names are X, Z, $AX(\Delta)$, $AZ(\Delta)$, $BX(\Delta,\Delta')$ and $BZ(\Delta,\Delta')$. The complex Z is the one we defined in chapter 3 using curve-systems, while $AZ(\Delta)$ is a generalization of the arc-system complex A we used in Chapter 2 to triangulate T_g^1. All the others are derived from these; in fact we have $X \subset Z$, $AX(\Delta) \subset AZ(\Delta)$ and $BX(\Delta,\Delta') \subset BZ(\Delta,\Delta') \subset AZ(\Delta)$.

i) <u>The</u> <u>Complexes</u> $AZ(\Delta)$ <u>and</u> $BZ(\Delta,\Delta')$

Let us be more specific. Fix $F = F_{g,r}^s$ and let Δ be a collection of points in ∂F. Also let Δ' be any proper subset of Δ. A <u>Δ-arc</u> will be the isotopy class of any C^∞ imbedded path in F from one point of Δ to another, or a C^∞ imbedded loop in F based at a point of Δ. The Δ-arc α is nontrivial if it is not null-homotopic and is not homotopic (rel $\partial\alpha$) into $\partial F - \Delta \cup \partial\alpha$.

A <u>Δ-arc system of rank k</u> is then any family $\alpha_0,\ldots,\alpha_k$ of non-trivial Δ-arcs which are disjoint, except that they may intersect at their endpoints, such that no two distinct α_i are homotopic (rel endpoints). Also define a <u>(Δ,Δ')-arc</u> to be any Δ-arc with one end in Δ' and the other in $\Delta - \Delta'$ and a <u>(Δ,Δ')-arc-system of rank k</u> to be any Δ-arc-system of rank k consisting only of (Δ,Δ')-arcs.

Now following the same pattern as for A let $AZ(\Delta)$ be the simplicial complex which has a k-cell $\langle\alpha_0,\ldots,\alpha_k\rangle$ for each rank k Δ-arc-system in F, with $\langle\alpha_0,\ldots,\alpha_k\rangle$ identified as a face of $\langle\beta_0,\ldots,\beta_\ell\rangle$ if and only if $\{\alpha_i\} \subset \{\beta_j\}$. Also define $BZ(\Delta,\Delta')$ to be the subcomplex of $AZ(\Delta)$ consisting of simplices $\langle\alpha_0,\ldots,\alpha_k\rangle$ where $\{\alpha_i\}$ is a (Δ,Δ')-arc-system.

We say that a simplicial complex $\sum$ of dimension n is <u>spherical</u> if $\sum \simeq \vee S^n$. The first step in the proof of Theorem 6.1 is:

<u>Theorem 6.3</u>: (a) The complex $AZ(\Delta)$ is contractible, except in the special cases where F is a 2-disk, a punctured 2-disk or an annulus with Δ contained in one component of ∂F, in which case $AZ(\Delta)$ is homeomorphic to a sphere.

(b) The complex $BZ(\Delta,\Delta')$ is spherical. If no component of ∂F contains points of both Δ' and $\Delta - \Delta'$, it is contractible.

We refer the reader to [H2] for the proof.

Along with these two complexes we may define $Z = Z^s_{g,r}$ just as we did Z^s_g; in fact $Z^s_{g,r} \cong Z^{r+s}_g$ because isotopy classes of simple closed curves in the interior of a surface aren't effected if we remove the boundary curves to create punctures. The mapping class group $\Gamma = \Gamma^s_{g,r}$ acts on each of the complexes Z, $AZ(\Delta)$ and $BZ(\Delta)$ in the obvious way.

ii) <u>The Complexes</u> X, $AX(\Delta)$, $BX(\Delta,\Delta')$

The problem with the complexes Z, $AZ(\Delta)$ and $BZ(\Delta,\Delta')$ is that their quotients by Γ, while finite complexes, are difficult to work with because they have too many cells. To rectify this problem we define in each case a natural subcomplex which is invariant under the action of Γ and has a simpler quotient.

The definition is the same in all 3 cases:

the subcomplexes X, $AX(\Delta)$ and $BX(\Delta,\Delta')$ are obtained by restricting in each case to curve or arcsystems which do not separate the surface (the entire system must be nonseparating, not just the individual curves or arcs). The complex X is $g-1$ dimensional since g curves whose union does not separate F will cut it into a planar surface where all further curves are separating. Similar reasoning shows that $AX(\Delta)$ and $BX(\Delta,\Delta')$ are $2g-2+r'$ dimensional where r' is the number of boundary components of F which contain points of Δ.

Theorem 6.4: The complexes X, $AX(\Delta)$ and $BX(\Delta,\Delta')$ are spherical.

This theorem is proved by a combinatorial analysis of the inclusion into the corresponding larger complex. For example, any map $f:S^n \to X$ with $n<g-1$ extends to $\overline{f}: D^{n+1} \to Z$ by Theorem 4.1. We then use the fact that n is small, so that simplices of $\mathrm{im}(\overline{f})$ do not involve too many curves in F, to show how to homotope $\overline{f}$ (rel f) into X. Similar arguments work for $AX(\Delta)$ and $BX(\Delta,\Delta')$ using an induction on g,r, $\#\Delta$ and $\#\Delta'$. This is why we had to allow Δ and Δ' to be arbitrary. From here on, however, we will assume that for $AX(\Delta)$, Δ consists of a single point and for $BX(\Delta,\Delta')$ Δ consists of 2 points on distinct boundary components of F with Δ' equal to one of the points. For brevity we then write $AX = AX(\Delta)$ and $BX = BX(\Delta,\Delta')$.

iii) Description of the Quotient Complexes

Let G be a group acting simplicially on the complex Σ. The quotient Σ/G inherits a natural cell structure from Σ only if the action has the property that whenever $g \in G$ fixes a simplex of Σ setwise, if does so pointwise. This property can always be arranged, if necessary, by passing to the first barycentric subdivision Σ^o of Σ. In our case this will only be necessary for X; the property above is already true for AX and BX.

The quotient of X^o by Γ is a simplex of dimension $g-1$. This is because any two rank-k curve systems which do not separate F are identified by Γ, so any top dimensional simplex of X^o maps homeomorphically to X^o/Γ.

The quotients of AX and BX are more complicated. Orient ∂F and consider any k-cell $\langle\alpha_0,\dots,\alpha_k\rangle$ of AX. In a neighborhood of the component of ∂F which contains the point q of Δ, the picture is as in Figure 6.1. Number the edges emanating from q in the order determined by the orientation; then $\langle\alpha_o,\dots,\alpha_k\rangle$ determines a pairing of the integers $\{1,\ldots,2k+2\}$.

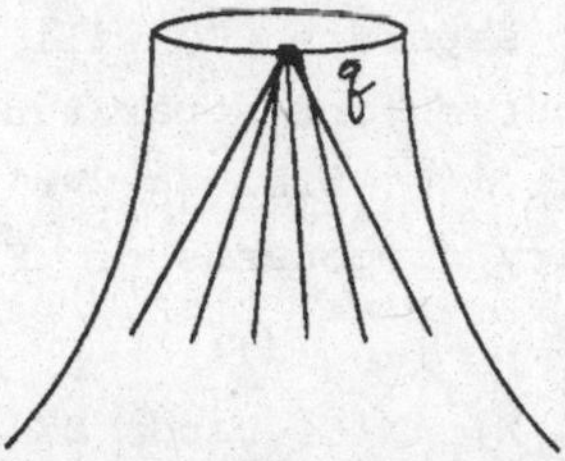

Figure 6.1

Two cells of AX are identified by Γ if and only if they give the same pairing. Furthermore, if $k < g$ any pairing may occur.

A similar analysis holds for BX. Number the edges emanating from q_1 as in figure 6.1; the order that they come into q_2 gives a permutation of $k+1$ elements. Any two cells of BX are identified by Γ if and only if they determine the same permutation and for $k \leq g$ any permutation can occur.

In the next section we will use these descriptions to show how stability is proven.

§ 2 Spectral Sequence Arguments

To prove Theorem 6.1 we use Theorem 6.4 and the spectral sequence techniques of [C], [Q], [W], [V]. Begin with the

Inductive assumptions;

(1_{k-1}) $\quad \phi_*: H_n(\Gamma_{g,r}) \to H_n(\Gamma_{g,r+1})$ is an isomorphism,

$g \geq 3n-2,\ 1 < n < k$ and $r \geq 1$.

$\phi_*: H_1(\Gamma_{g,r}) \to H_1(\Gamma_{g,r+1})$ is an isomorphism,

$g \geq 2,\ r \geq 1$.

(2_{k-1}) $\quad \psi_*: H_n(\Gamma_{g,r}) \to H_n(\Gamma_{g+1,r-1})$ is surjective,

$g \geq 3n-2,\ 1 \leq n < k,\ r \geq 2$, and an isomorphism,

$g \geq 3n-1,\ 1 < n < k,\ r \geq 2$.

$\psi_*: H_1(\Gamma_{g,r}) \to H_1(\Gamma_{g+1,r-1})$ is an isomorphism,

$g \geq 3,\ r \geq 2$.

(3_{k-1}) $\eta_*: H_n(\Gamma_{g,r};\mathbb{Q}) \to H_n(\Gamma_{g+1,r-2};\mathbb{Q})$ is an isomorphism,

$g \geq 3n-1,\ n < k,\ r \geq 2.$

$\eta_*: H_n(\Gamma_{g,r}) \to H_n(\Gamma_{g+1,r-2})$ is an isomorphism,

$g \geq 3n,\ n < k,\ r \geq 2.$

Both maps are surjective at one smaller value of g.

It is easy to check (1_k), (2_k) and (3_k) directly for $k \leq 2$ using the results of [H1] (see chapter 7). Assume $k \geq 3$ and that (1_j), (2_j) and (3_j) have been verified for all $j < k$. There are then three spectral sequence arguments to perform to prove (1_k), (2_k) and (3_k). We will show how (1_k) goes (it is the easiest); the reader is refered to [H2] for the other two.

First notice that ϕ_* is injective for every g, k when $r \geq 1$. This is because we may define $\theta: \Gamma_{g,r+1} \to \Gamma_{g,r}$ by plugging one of the extra boundary components of the pair of pants sewed on in defining ϕ and we will have $\theta \circ \phi = 1$.

Now look at the action of $\Gamma = \Gamma_{g,r+1}$ on the complex BX defined on $F_{g,r+1}$. Let $K_* \to \mathbb{Z}$ be the augmented chain complex of BX. Choose $E_* \to \mathbb{Z}$, a free $\mathbb{Z}\Gamma$ resolution of $\mathbb{Z}$. The double complex $E_* \otimes_{\mathbb{Z}\Gamma} K_*$ gives rise to a spectral sequence converging to zero for $p+q \leq 2g-1$, with

$$E^1_{p,q} = H_q(\Gamma;K_p), \quad p \geq -1,$$

(see [C], [Q], [W], [V]). By Shapiro's lemma there is a decomposition

$$H_q(\Gamma;K_p) \cong \bigoplus_{i=1}^{n_p} H_q(\Gamma^i_p)$$

Where $K_p = K^1_p \oplus \ldots \oplus K^{n_p}_p$ with $\Gamma(K^i_p) \subset K^i_p$, Γ acts transtively on the generators of each K^i_p and Γ^i_p denotes the stabilizer of a chosen p-cell σ^i_p, a generator from K^i_p.

<u>Lemma</u> <u>6.5</u>: $H_p(BX/\Gamma) = 0,\ 1 \leq p \leq g-1.$

<u>Proof</u>: Recall each p-cell σ^i_p may be defined by the permutation it gives of $0,1,\ldots,p$. Label this permutation $(i_0,\ldots,i_p)$ where i_j is the image of j under the permutation. The boundary map of BX/Γ is now

$$\partial_p(i_0,\ldots,i_p) = \sum_{j=0}^{p} (-1)^j(\tau_j(i_0),\ldots,\hat{i}_j,\ldots,\tau_j(i_p))$$

where τ_j is the shift map

$$\tau_j(n) = \begin{cases} n, & n < i_j \\ n-1, & n > i_j. \end{cases}$$

Consider the map

$$C_p(BX/\Gamma) \xrightarrow{D_p} C_{p+1}(BX/\Gamma)$$

$$(i_0,\ldots,i_p) \to (p+1,i_0,\ldots,i_p).$$

The formula $D\partial + \partial D = 1$ is easy to verify, so the identity and the zero map on $C_*(BX/\Gamma)$ are chain homotopic in the range $p < g$. This proves 6.5. □

Next we look at the E^1 term of the spectral sequence. If necessary, rechoose the $\sigma_p^i = \langle\beta_0^i,\ldots,\beta_p^i\rangle$ so that for each p there is an embedding

$$\omega_p : F_{g-p,1} \to F - \{\beta_t^i ; 0 \le t \le p, 1 \le i \le n_p\},$$

with $\omega_p = \omega_{p-1} \cdot \psi \cdot \phi$ for each p. This is possible because the genus of $F - \{\beta_j^i ; 0 \le j \le p\}$ is at least $g - p$. If $(\sigma_p^j)_i$ is the ith face of σ_p^j, choose f_i in Γ identifying $(\sigma_p^j)_i$ with the appropriate $\sigma_{p-1}^{k_i}$, making sure that $f_i|\omega_p(F_{g-p,1}) = 1$. The map

$$d^1 : E^1_{p,q} \to E^1_{p-1,q}$$

has components

$$(-1)^i (f_i)_* : H_q(\Gamma_p^j) \to H_q(\Gamma_{p-1}^{k_i}).$$

By (1_{k-1}) and (2_{k-1}), the maps $(\omega_{p-1})_*$, $(\omega_p)_*$, ψ_* and ϕ_* are isomorphisms as long as $g - p \ge 3q$, $q < k$, and are surjective when $g-p = 3q-1$, $q < k$. Since

$$(\omega_{p-1})_* \cdot \psi_* \cdot \phi_* = (f_i)_*(\omega_p)_*,$$

$(f_i)_*$ is an isomorphism or is surjective for the same values of p, q and g.

Consider the diagram

$$\begin{array}{ccc} H_q(\Gamma^i_p) & \rightarrow & H_q(\Gamma) \\ \downarrow{\scriptstyle (f_i)_*} & & \downarrow{\scriptstyle (f_i)_*} \\ H_q(\Gamma^{k_i}_{p-1}) & \rightarrow & H_q(\Gamma) \end{array} .$$

The horizontal maps are induced by the inclusions of the stabilizers in Γ. Since F_i is an element of Γ, it induces the identity on homology, so that the right hand vertical map is the identity. The horizontal maps are compositions of the maps ϕ_* and ψ_*; so for $p \geq 1$ and $q < k$, they are isomorphism. This means that $E^1_{*,q}$ is isomorphic as a complex to $C_*(BX/\Gamma) \otimes H_q(\Gamma)$, where the boundary map is the tensor product of the boundary map of BX/Γ with the identity on $H_q(\Gamma)$. In particular,

$$E^2_{p,k-p} \cong H_p(BX/\Gamma; H_{k-p}(\Gamma)) = 0,$$

as long as $g - (p+1) \geq 3(k-p) - 1$. Since $p \geq 1$, this holds for $g \geq 3k-2$.

Now we use the fact that $E^\infty_{p,q} = 0$, $p+q<2g$, to conclude that

$$d^1_{0,k} : E^1_{0,k} \rightarrow E^1_{-1,k}$$

is surjective, $g \geq 3k-2$ ($k>1$ and $g \geq 3k-2$ imply $g>k$). As $d^1_{0,k}$ is the map ϕ_*, (1_k) is verified.

The arguments for (2_k) and (3_k) are more difficult and will be omitted.

Chapter 7. Computations of the Cohomology of Γ_g^s

Finally the time has come to construct some cohomology classes for Γ_g^s. In this chapter we will discuss which Betti numbers of Γ_g^s and $\mathcal{M}_g^s$ are known explicitly. Then we will give Miller's construction of a polynomial algebra on even generators in the stable cohomology of Γ_g^s [Mi], and describe unstable relations for these classes due to Mumford [Ml], Morita [Mor] and Harris [Ha].

§1. The first two Betti numbers of M_g^s

In Theorem 3.2 we observed that Γ_g^s has no rational cohomology above dimension $4g-4+s$ $(4g-5$ when $g > 1, s = 0)$, while M_g^s has no integral homology above dimension $4g-4+s$, $s > 0$. Only two other Betti numbers, β_1 and β_2, are known. Mumford [M] proved that $H_1(\Gamma_g)$ is torsion of order dividing 10 $(g > 1)$ and Powell [P] proved that $H_1(\Gamma_g) = 0$ when $g > 2$. The general statement is:

Theorem 7.1:

$$H_1(\Gamma_g^s) \cong \begin{cases} 0 & \text{for } g > 2, \\ \mathbb{Z}/10\mathbb{Z} & \text{for } g = 2 \text{ and} \\ \mathbb{Z}/12\mathbb{Z} & \text{for } g = 1. \end{cases}$$

Another thing Mumford did in [M] was to define $\mathrm{Pic}(M_g^s)$ and prove $\mathrm{Pic}(M_g^s) \cong H^2(\Gamma_g^s)$, $g > 1$. In [Hl] this latter group was computed; the result is:

Theorem 7.2: $H_2(\Gamma_g^s) \cong \mathbb{Z}^{s+1}$ for $g > 4$.*

This result also holds rationally for Γ_3^s and Γ_4^s while $H_2(\Gamma_2^s;\mathbb{Q})$ and $H_2(\Gamma_1^s;\mathbb{Q})$ have rank s. We do not know the torsion in $H_2(\Gamma_g^s)$ when $2 \le g \le 4$.

The proofs of Theorem 7.1 and 7.2 will not be presented here as that would lead us too far afield. Instead we content ourselves with a description of generators for $H_2(\Gamma_g^s)$. Geometrically, we saw in Chapter 5 that $\frac{1}{\pi^2}\omega_{WP}$ generates $H^2(M_g;\mathbb{Q})$. We now give a purely topological description of generators. Let $\Lambda = \mathrm{Diff}^+(F_g^s)$ be the group of orientation preserving diffeomorphism of F_g which fix s points, equipped with the C^∞-topology. The group Λ has one

---

* Actually, the statement in $[H_l]$ was that $H_2(\Gamma_g^0)$ has torsion, but this was in error, the correction appears in the same journal somewhat later.

component for each element of Γ_g^s and, when $2-2g-s<0$, Earle and Eells [EE] have proved that each of these components is contractible. This means that the classifying space $B\Lambda$ is a $K(\Gamma_g^s,1)$.

Let $\Omega_2(B\Lambda)$ denote the second bordism group of $B\Lambda$. An element of $\Omega_2(B\Lambda)$ is represented by a map $\phi : X \to B\Lambda$ where X is a closed oriented surface and two maps $\phi_1 : X_1 \to B\Lambda$ and $\phi_2 : X_2 \to B\Lambda$ represent the same element if and only if there is a map $\psi : M^3 \to B\Lambda$ where M^3 is an oriented 3-manifold with $\partial M^3 = X_1 \coprod - X_2$ and $\psi | X_i = \phi_i$. Bordism groups and homology groups agree in low dimensions; in particular we have:

$$H_2(\Gamma_g^s) \cong H_2(B\Lambda) \cong \Omega_2(B\Lambda).$$

Any map $f : X \to B\Lambda$ induces a bundle over X with fiber F: use f to pull back the bundle $E\Lambda \times_\Gamma F \to B\Lambda$ where $E\Lambda$ is the universal cover of $B\Lambda$. Bordant maps pull back bordant bundles, i.e. bundles which cobound a bundle over a 3-manifold.

Now we may define $H_2(\Gamma_g^s) \to \mathbb{Z}^{s+1}$. Each $X \in H_2(\Gamma_g^s)$ gives rise as above to a fiber bundle $F_g \to W^4 \to X$ with structure group in Λ. This bundle then has s canonical sections $\sigma_1, \ldots, \sigma_s$; the invariants we associate to X are signature$(W)/4$ and the self-intersection numbers $[\sigma_i(X)]^2$ (the signature of W^4 is always divisible by 4 so these invariants are all integers). Theorem 7.2 states that this map is an isomorphism.

§2. Miller's Polynomial Algebra

Let $\Lambda = \mathrm{Diff}^+(F_g^0)$ and consider the universal F_g bundle $p : E \to B\Lambda$. Define τ to be the bundle of tangents to the fibers of p and set $\omega = -c_1(\tau)$. Now define

$$\kappa_i = p_* \omega^{i+1};$$

$\kappa_i \in H^{2i}(B\Lambda) \cong H^{2i}(\Gamma_g^0)$. When we have an analytic family $\pi : X \to B$ which coincides with the pull back of $E \to B\Lambda$:

$$\begin{array}{ccc} X & \xrightarrow{\hat{f}} & E \\ \downarrow & & \downarrow \\ B & \xrightarrow{f} & B\Lambda, \end{array}$$

then $\hat{f}^*(\omega)$ is c_1 of the relative dualizing sheaf. Mumford [M1] shows that $f^*(\kappa_i)$ may be regarded as a Chow cohomology class in $A^i(B) \otimes \mathbb{Q}$. Miller proves:

Theorem 7.3: Let $\mathbb{Q}[\kappa_1,\kappa_2,\ldots]$ be the polynomial algebra on the κ_i. Then the induced map

$$\mathbb{Q}[\kappa_1,\kappa_2,\ldots] \to H^*(\Gamma_g^0;\mathbb{Q})$$

is injective in dimensions less than $g/3$.

To prove this theorem, Miller first looks at the Hopf algebra $A = \varinjlim H_*(\Gamma_{g,1};\mathbb{Q})$ defined in Chapter 6 and proves that each κ_i vanishes on the λ_* decomposibles. It therefore suffices to construct explicit homology classes $X_n \in H_{2n}(\Lambda;\mathbb{Q})$ such that $[\kappa_n, X_n] \neq 0$. Theorem 7.3 then follows from the stability Theorem 6.1.

i) Construction of the Classes X_n

The classes X_n are provided by:

Lemma 7.4: There exists for each $n > 0$ a smooth fibration of projective algebraic varieties $\pi : X^{n+1} \to B^n$ with fiber F_g such that $[\omega^{n+1}, X] \neq 0$, where ω is c_1 of the tangents to the fiber of π.

The construction of the fibration $\pi : X^{n+1} \to B^n$ is modeled on one of Atiyah [A] . First we review Atiyah's construction which gives the example for $n = 1$. Let C be a connected curve of genus $g > 2$ and let $\tau : C \to C$ be a free involution. The base B^1 is the 2^{2g}-fold covering of C determined by the map $\pi_1(C) \to H_1(C;\mathbb{Z}/2\mathbb{Z})$; if $f : B^1 \to C$ is the covering map, then f^* is 0 on H^1 with $\mathbb{Z}/2\mathbb{Z}$-coefficients. Now look at $B^1 \times C$ and consider the graphs G_f and $G_{\tau f}$. Atiyah's choice of f guarantees that the homology class of $G_f + G_{\tau f}$ in $H_2(B^1 \times C;\mathbb{Z})$ is even, therefore it is possible to form the 2-fold cover $X^2 \to B^1 \times C$ ramified along the divisor $G_f + G_{\tau f}$. Composing with the projection to B^1 gives $\pi : X^2 \to B^1$ with fiber of genus $2g$. Atiyah now shows that

$$\omega^2/3 \;=\; \text{signature}\,(X^2) \;=\; (g-1)2^{2g-1}$$

where ω is c_1 of the tangents to the fibers.

To generalize the construction, fix an epimorphism $H_1(C;\mathbb{Z}) \to \mathbb{Z} \oplus \mathbb{Z}$; the composition $\pi_1(C) \to H_1(C;\mathbb{Z}) \to \mathbb{Z} \oplus \mathbb{Z} \to \mathbb{Z}/2^n\mathbb{Z} \oplus \mathbb{Z}/2^n\mathbb{Z}$ has kernel K_n and determines a 4^n-fold covering $\dot{C}_n \to C$. Miller uses the tower $\to C_n \to \ldots \to C_1 \to C_0 = C$ to inductively define $X^{n+1} \to B^n$ with connected fiber Y_n such that:

(1) X^{n+1} maps onto X^n in such a way that the composition

$$X^{n+1} \to X^n \to \dots \to X^2 \to B^1 \times C \to C$$

maps $\pi_1(Y^n)$ and $\pi_1(X^{n+1})$ onto K_{n-1}, and

(2) $[\omega^{n+1}, X^{n+1}] \neq 0$, $\omega = c_1$ (tangents to the fiber) as above.

Assume inductively that $X^{i+1} \to B^i$ has been constructed for all $i < n$. By (1), $X^n \to C$ lifts to $X^n \to C_{n-2}$ and sends both $\pi_1(Y_{n-1})$ and $\pi_1(X^n)$ onto K_{n-2}. Define X' to be the 4-fold cover $X' \to X^n$ induced by the composition $\pi_1(X^n) \to K_{n-2} \to K_{n-2}/K_{n-1} \cong \mathbb{Z}/2\mathbb{Z} \oplus \mathbb{Z}/2\mathbb{Z}$, with a similar definition for $Y' \to Y_{n-1}$. Clearly both $\pi_1(X') \to \pi_1(C)$ and $\pi_1(Y') \to \pi_1(C)$ have image K_{n-1}. Form the fiber product

$$\begin{array}{ccc} X' \times_{B^{n-1}} X' & \xrightarrow{p_2} & X' \\ {\scriptstyle p_1}\downarrow & & \downarrow \\ X' & \longrightarrow & B^{n-1} \end{array} ;$$

the fiber of p_1 is Y'. We extend this diagram to the left as follows. Let $T^n \to X'$ be the finite covering determined by the kernel of the representation $\pi_1(X') \to \mathrm{Aut}(H_1(Y';\mathbb{Z}/2\mathbb{Z}))$ associated to p_1. Also let $D^n \to T^n$ be the finite covering determined by $\pi_1(T^n) \to H_1(T^n; \mathbb{Z}/2\mathbb{Z})$. Then we have the pull back diagram

$$\begin{array}{ccccc} W^{n+1} & \xrightarrow{h} & & & X' \times_{B^{n-1}} X' \\ \downarrow & & & & \downarrow{\scriptstyle p_1} \\ B^n & \longrightarrow & T^n & \longrightarrow & X' \end{array} .$$

Now $X' \times_{B^{n-1}} X'$ admits the free involution $\sigma(x_1,x_2) = (x_1,\tau x_2)$ where τ is a nontrivial covering translation of $X' \to X$. If Δ denotes the diagonal section $\Delta(x) = (x,x)$ of p_1, then we may look at the divisor $\Delta + \sigma\Delta$ in $X' \times_{B^{n-1}} X'$. Miller proves that the homology class of $h^{-1}(\Delta+\sigma\Delta)$ is even. Therefore it makes sense to define X^{n+1} to be the 2-fold cover of W^{n+1} ramified along $h^{-1}(\Delta+\sigma\Delta)$. The composition $X^{n+1} \to W^{n+1} \to B^n$ is the desired fibration.

The entire construction is summarized in the diagram

$$\begin{array}{ccccccc}
X^{n+1} & & & & & & \\
\downarrow & & & & & & \\
W^{n+1} & \xrightarrow{h} & X' \times X' & \xrightarrow{p_2} & X' & \longrightarrow & X^n \\
\downarrow & & \downarrow{\scriptstyle p_1} & & \downarrow & & \downarrow \\
B^n \longrightarrow T^n & \longrightarrow & X' & \longrightarrow & B^{n-1} & = & B^{n-1}.
\end{array}$$

The map $p : X^{n+1} \to X^n$ promised in (1) above is the obvious one in this diagram; it is easy to check it has the right properties.

To calculate $[\omega^{n+1}, X^{n+1}]$ choose a holomorphic differential on C and use the maps of (1) to pull it back to ω_{n+1} on X^{n+1} and ω_n on X^n. One computes that

$$\omega_{n+1} = p^*(\omega_n) - (h^{-1}(\Delta+\sigma\Delta))$$

and then uses this to work back through the diagrams to see

$$[\omega_{n+1}^{n+1}, X^{n+1}] = 16N((1/2)^{n+1}-1)[\omega_n^n, X^n] \neq 0$$

where N is the degree of the covering $B^n \to X'$ (see [Mi] for details)

ii) <u>Comparison with</u> $H^*(Sp(2g;\mathbb{Z});\mathbb{Q})$

In [B] Borel found the stable cohomology of $Sp(2g;\mathbb{Z})$ with $\mathbb{Q}$-coefficients; it is a polynomial algebra on generators in dimensions $4k+2$, $k \geq 0$. Miller considers the problem of computing the map induced on cohomology by $\mu : \Gamma_g \to Sp(2g;\mathbb{Z})$.

The map μ induces a map $BDiff^+(F_g) \to BSp(2g;\mathbb{Z})$. Using the inclusion $Sp(2g;\mathbb{Z}) \subset Sp(2g;\mathbb{R})$ and the fact that $U(g)$ is the maximal compact subgroup of $Sp(2g;\mathbb{R})$ we obtain a map

$$\bar{\mu} : BDiff^+(F_g) \to BU(g).$$

The map $\bar{\mu}$ sends the λ product in A to the product in BU induced by Whitney sum. If ν denotes the universal bundle over BU, the characteristic classes $S_n(\nu)$ (the polynomials in the Chern classes of ν corresponding to $\sum t_j^n$) vanish on decomposibles. Miller [Mi], Mumford [M1] and Morita [Mor] prove independently that

$$\bar{\mu}^*(s_n(\nu)) = \begin{cases} 0 & n \text{ even}, \\ (-1)^{\frac{n+1}{2}} \dfrac{B_{n+1}}{n+1} \cdot \kappa_n & n \text{ odd}, \end{cases}$$

where B_n is the $n^{\underline{th}}$ Bernoulli number. In particular, this implies μ^* is injective.

iii) Relations among the κ_i

In [Ml], Mumford gave an algebraic construction of the classes κ_i and proved relations between them. These and other relations (along with Miller's polynomial algebra) were rediscovered by Morita [Mor] who described them topologically. Harris ([Ha]), while considering the problem of determining when certain linear combinations of divisor classes are ample and/or effective, also find relations among the classes in $\mathrm{Pic}(M_g^1) \otimes \mathbb{Q}$. We will briefly describe these relations.

Let η be the g-dimensional complex vector bundle over $B\Lambda$ ($\Lambda = \mathrm{Diff}^+ F_g$ as before) induced by $\bar{\mu}$, and let $s_i(\eta)$ be the pull backs of the characteristic classes $s_i(\nu)$ by $\bar{\mu}$. As a real bundle, η is determined by the map $\Gamma_g \to Sp(2g;\mathbb{R})$, so it is flat. This means all its Pontrjagin classes vanish, or equivalently

$$s_{2i}(\eta) = 0 \qquad \text{for all } i. \tag{R1}$$

This is a relation in $H^{4i}(\Gamma_g^0;\mathbb{Q})$.

Let $\xi = \pi\colon E \to B\Lambda$ be the universal bundle over $B\Lambda$ and consider the conjugate of the pull back $\overline{\pi^*(\eta)}$ over E. The fibers of the bundle η over the point x may be identified naturally with $H^1(E_x;\mathbb{R})$ which in turn may be identified with the space of harmonic 1-forms on E_x. Using this description, define a map $\phi\colon \overline{\pi^*(\eta)} \to \xi^*$ by setting

$$\phi(\omega)(v) = \omega(v) + i\omega(v).$$

This gives a sequence

$$0 \to \mathrm{Ker}(\phi) \to \overline{\pi^*(\eta)} \to \xi^* \to 0;$$

Mumford and Morita observe $c_k(\mathrm{Ker}\ \phi) = 0$ for $k \geq g$. This gives

$$\sum_{j=0}^{g} \kappa^{k-j} c_j = 0, \quad k \geq g, \tag{R2}$$

where $\kappa = \pi^*(\kappa_1) \in H^2(E)$ and c_j denotes the pull back to $H^{2j}(E)$ of the $j^{\underline{th}}$ Chern class of η. Since E is a $K(\Gamma_g^1,1)$, this is a relation in $H^*(\Gamma_g^1;\mathbb{Q})$.

Applying π_* to (R2) gives

$$\sum_{j=0}^{g} \kappa_{k-1-j}\, \pi_*(c_j) = 0, \quad k \geq g, \tag{R3}$$

a relation in $H^*(\Gamma_g;\mathbb{Q})$. Mumford shows that these relations imply that κ_k is a polynomial in $\kappa_1,\ldots,\kappa_{g-2}$ for all $k \geq g-1$.

In [Ha] Harris considers the following question. Let ω be c_1 of the relative dualizing sheaf as before and let $\kappa = \pi^*\pi_*\omega^2 \in H^2(M_g;\mathbb{Q}) \cong \mathrm{Pic}(M_g) \otimes \mathbb{Q}$. Consider complete, nondegenerate families of curves $X \to B$ with associated classifying map $\phi : B \to M_g$ (non-degenerate means ϕ has finite fibers). Then for what a, b is $a\omega + b\kappa$ ample on every $X \to B$? He proves this is true for $a > 0$ and $a + 4g(g-1)b > 0$. He then shows that if B is 2-dimensional,

$$\kappa_1^2 > (2g-2)\kappa_2 \qquad \text{(R4)}$$

and finally, the most amazing relation of all is

$$(4g(g-1)\omega - \kappa)^{g+1} = 0. \qquad \text{(R5)}$$

We refer the reader to [Ha] for more information and proofs.

§3. Homology of $\overline{M}_g^s$

Using Mayer-Vietoris one can easily relate $H_*(M_g^s)$ to $H_*(\overline{M}_g^s)$. For example, $\beta_1(M_g^s) = 0$ implies $\beta_1(\overline{M}_g^s) = 0$ and we saw in Chapter 5 that $\beta_2(M_g) = 1$ implies $\beta_2(\overline{M}_g) = 2 + [g/2]$. The fact that κ_1 is symplectic on $\overline{M}_g$ shows that $\kappa_1^{3g-3} \neq 0$. What is more, Wolpert uses intersections of the D_i to show that $\beta_{2k}(\overline{M}_g) \geq \frac{1}{2}\binom{g-1}{k}$.

Let $\mathfrak{h}_g$ denote the Siegel upper half space of degree g,

$$\mathfrak{h}_g = \{Z \in M_g(\mathbb{C}) : Z = Z^t, \operatorname{Im} Z > 0\}$$

where $M_g(\mathbb{C})$ is $g \times g$ matrices with complex entries and t denotes transpose. The group $Sp(2g,\mathbb{Z})$ acts on $\mathfrak{h}_g$ by the formula

$$Z \cdot M = (ZC+D)^{-1} \cdot (ZA+B)$$

where $M = \begin{pmatrix} A & B \\ C & D \end{pmatrix}$. The quotient $\mathfrak{h}_g/Sp(2g;\mathbb{Z})$ is called the <u>Siegel modular space</u> and may be identified with A_g, the space of <u>principally polarized abelian varieties</u>. The action of $Sp(2g;\mathbb{Z})$ is properly discontinuous and $\mathfrak{h}_g \simeq *$ so $H^*(A_g;\mathbb{Q}) \cong H^*(Sp(2g;\mathbb{Z});\mathbb{Q})$.

One may associate to each Reimann surface its Jacobian; this defines a mapping

$$J : M_g \to A_g$$

called the <u>period mapping</u>. The classical Torelli theorem says that J is an embedding. Rationally, the maps J^* and μ^* are the same on cohomology.

Suppose next that $\overline{A}_g$ denotes the Satake compactification of A_g;

it is obtained in a manner similar to $\overline{M}_g$ by adding on copies of A_k, $k < n$, at ∞. The map J extends to $\overline{J}:\overline{M}_g \to \overline{A}_g$ by normalizing and then taking period matrices as before. Charney and Lee ([CL1]) prove three things in this situation which we summarize as:

Theorem 7.5: (1) For $g \geq i+1$, $H^i(\overline{A}_g;\mathbb{Q}) \cong H^i(\varinjlim_g \overline{A}_g;\mathbb{Q})$.

(2) $H^*(\varinjlim \overline{A}_g;\mathbb{Q}) \cong \mathbb{Q}[x_2,x_6,\ldots] \otimes \mathbb{Q}[y_6,y_{10},\ldots]$, where the x_i live in $H^*(A_g;\mathbb{Q})$.

(3) $\mathrm{Ker}(\overline{J}^*) = \langle y_6,y_{10},\ldots\rangle$; that is, image $(\overline{J}^*)$ = image (J^*) = image(μ^*).

This result is proven by decomposing $\overline{A}_g$ as a union of simplicial $K(\pi,1)$'s, relating the pieces to K-theory using $\mathbb{Q}$ categories and then applying results from K-theory to compute the cohomology.

§4. Torsion in $H^*(\Gamma_g;\mathbb{Z})$

For completeness we will state here what is known about torsion in $H^*(\Gamma_g^0;\mathbb{Z})$. By stability, $H^*(\Gamma_g^0;\mathbb{Z}) \cong H^*(\Gamma_{g,1}^0;\mathbb{Z})$ and since $\Gamma_{g,1}^0$ is torsion free, $H^*(\Gamma_{g,1}^0;\mathbb{Z}) \cong H^*(M_{g,1}^0;\mathbb{Z})$ where $M_{g,1}^0$ is the moduli space of triples (R,p,v) with R a Riemann surface, p a point of R and v a unit tangent vector to R at p. However, we only have $H^*(\Gamma_g;\mathbb{Q}) \cong H^*(M_g;\mathbb{Q})$; it is not known if the torsion classes we shall describe lie in M_g.

Outside of low dimensional phenomena, the first construction of torsion in $H^*(\Gamma_g)$ seems to be due to Charney and Lee [CL]. As a special case of their work on classifying spaces of Hodge structures, they prove that for every odd prime p there exists a p-torsion class in $H^{2p-2}(\Gamma^0_{(p-1)/2};\mathbb{Z})$. This result was strengthened by Glover and Mislin [GM] who showed:

Theorem 7.6: The stable cohomology group $H^{4k}(\Gamma_g;\mathbb{Z})$ $(g \gg k)$ contains an element of order E_{2k} = denominator of $B_{2k}/2k$, where B_{2k} is the $2k^{\text{th}}$ Bernoulli number.

For p prime, the number E_{2k} is divisible by p^α if and only if $p^{\alpha-1}(p-1)$ divides $2k$. Together with Theorem 7.6, this can be used to show

Corollary 7.7: Let p be prime and odd and let $\alpha \geq 1$. Then, for every $j \geq 1$, $H^{2jp^\alpha(p-1)}(\Gamma_g^0;\mathbb{Z})$ contains an element of order p^α.

Chapter 8: The Euler Characteristic of Moduli Space

The results in this chapter are all taken from [HZ].

Let $\hat{\Gamma}$ be a torsion free subgroup of finite index in Γ_g^s. The subgroup $\hat{\Gamma}$ acts properly discontinuously and freely on T_g^s, the quotient is therefore a manifold. We define the orbifold Euler characteristic of Γ_g^s to be

$$\chi(\Gamma_g^s) = [\Gamma_g^s : \hat{\Gamma}]^{-1} \cdot e(T_g^s/\hat{\Gamma})$$

where $e(T_g^s/\hat{\Gamma})$ is the ordinary Euler characteristic of $T_g^s/\hat{\Gamma}$. The rational number $\chi(\Gamma_g^s)$ does not depend on the choice of the subgroup $\hat{\Gamma}$.

We will use the contractible cell complex Y of Chapter 3 to compute $\chi(\Gamma_g^s)$. Using the exact sequence (S_1) of Chapter 1 together with the fact that orbifold Euler characteristics are multiplicative in short exact sequences, we see that it suffices once again to consider the case where $s = 1$. The main theorem of [HZ] is:

Theorem 8.1: $\chi(\Gamma_g^1) = \zeta(1-2g)$.

Here ζ denotes the Riemann zeta function; its value at $1-2g$ is given by $\zeta(1-2g) = -B_{2g}/2g$, where B_{2g} is the $2g^{\underline{th}}$ Bernoulli number.

An equivalent formulation of the theorem is given by

$$\chi(\Gamma_g) = \frac{B_{2g}}{4g(g-1)} \quad (g > 1),$$

and the general expression is

$$\chi(\Gamma_0^s) = \begin{cases} 1 & s \leq 3 \\ (-1)^{s-3}(s-3)! & s \geq 3 \end{cases}$$

$$\chi(\Gamma_1^s) = \begin{cases} -\frac{1}{12} & s \leq 1 \\ \frac{(-1)^s (s-1)!}{12} & s \geq 1 \end{cases}$$

$$\chi(\Gamma_g^s) = (-1)^s \frac{(2g+s-3)!}{2g(2g-2)!} B_{2g} \qquad g \geq 2, \quad s \geq 0.$$

By a result of Ken Brown ([Br]), the ordinary Euler characteristic

$$e(\Gamma_g^s) = \sum_{i\geq 0} (-1)^i \text{ rank } (H_i(\Gamma_g^s;\mathbb{Q}))$$

can be computed from the orbifold Euler characteristics of the centralizers of the elements of finite order in Γ_g^s. The exact formula is

$$e(\Gamma_g^s) = \sum_{<\sigma>} \chi(Z_\sigma)$$

where the sum is over all conjugacy classes $<\sigma>$ of elements σ of finite order in Γ_g^s and Z_σ denotes the centralizer of σ. These centralizers are extensions of finite groups by mapping class groups, so their Euler characteristics can be computed from the formulas above. Working this out for $s = 0$ gives:

Theorem 8.2: The numbers $e(\Gamma_g)$ are given by the generating function

$$\sum_{g\geq 1} e(\Gamma_g)t^{2g-2} = \sum_{\substack{k\geq 1\\ m,d|k}} \sum_{\substack{h,s\geq 0\\ s+2h\geq 3}} \frac{\chi(\Gamma_h^s)}{s!} \frac{\mu(m)}{m^2} \phi(d) (\tfrac{k}{m}t^k)^{2h-2} \beta_{\frac{k}{m},\frac{d}{(d,m)}} (t^m)^s,$$

where μ denotes the Möbius function, ϕ the Euler totient function and β is given by

$$\beta_{k,d}(t) = \sum_{\substack{\ell | k\\ \ell\neq k}} \mu(\frac{d}{(d,\ell)}) \frac{\phi(k/\ell)}{\phi(d/(d,\ell))} t^{k-\ell}.$$

This theorem can be used to easily derive values of $e(\Gamma_g)$. For example, we have the following table when $g \leq 15$:

g	$e(\Gamma_g)$
1	1
2	1
3	3
4	2
5	3
6	4
7	1
8	-6
9	45
10	-86
11	173
12	-100
13	2641
14	-48311
15	717766

An important consequence of Theorem 8.2 is that it can be used to show $e(\Gamma_g) \sim \chi(\Gamma_g)$; therefore the Betti numbers of Γ_g grow more that exponentially in g and Γ_g has a lot of homology in dimensions congruent to $g-1$ modulo 2. We have seen how to construct even dimensional classes in Chapter 7 (but not enough to account for the size of $e(\Gamma_g)$ even when g is odd); we know of no constructions of odd-dimensional classes.

§1. Basics of the Proof

There are three parts to the proof of Theorem 8.1; we outline them now. Let P_n be a polygon with $2n$ sides; a pairing of the sides of P_n gives a unique oriented surface of genus $g \leq n/2$. Using these pairings we define three double sequences of numbers: $\varepsilon_g(n)$ is the number of pairings of the edges of P_n which give a surface of genus g, $\mu_g(n)$ is the number of parings of the edges of P_n which give a surface of genus g, but with no edge paired to its neighbor (figure 8.1 configuration A) and $\lambda_g(n)$ is the number of pairings giving a surface of genus g, but without an occurence of either configuration A or B of figure 8.1.

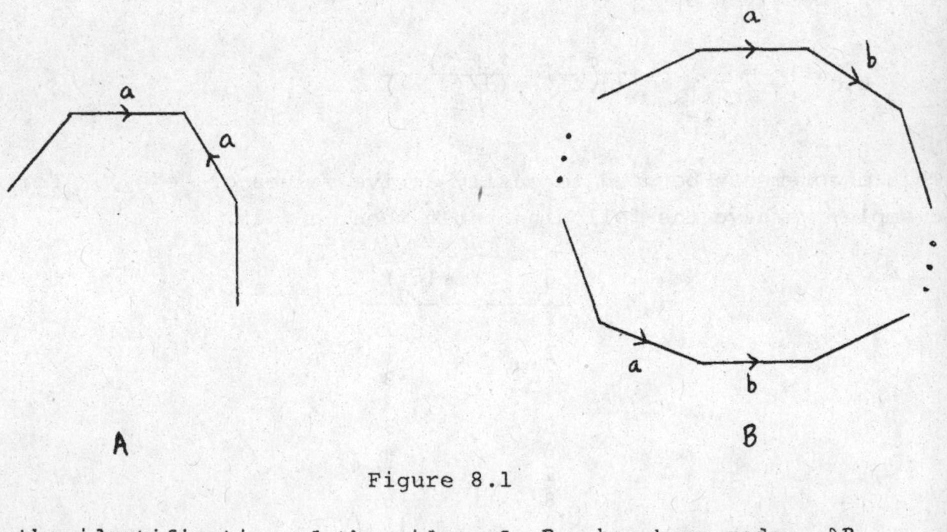

Figure 8.1

After the identification of the sides of P_n has been made, ∂P_n becomes a finite graph $\Omega \subset F$ (compare Chapter 2). The conditions above say that $\varepsilon_g(n)$ counts everything, but $\mu_g(n)$ counts only the pairings for which every vertex of Ω has valence ≥ 2 and $\lambda_g(n)$ counts pairings Ω with vertices of valence ≥ 3.

By analyzing the action of Γ_g^1 on Y we will prove in §2:

<u>Theorem 8.3</u>: $\chi(\Gamma_g^1) = \sum_{n=2g}^{6g-3} (-1)^{n-1} \lambda_g(n)/2n.$

It is not difficult to relate the numbers $\varepsilon_g(n)$, $\mu_g(n)$ and $\lambda_g(n)$. The relation is provided by the following lemma whose proof will be omitted

<u>Lemma 8.4</u>: (a) $\varepsilon_g(n) = \sum_{i\geq 0} \binom{2n}{i} \mu_g(n-i)$

(b) $\mu_g(n) = \sum_{i\geq 0} \binom{n}{i} \lambda_g(n-i)$.

In §3 we will give the main combinatorial result:

<u>Theorem 8.5</u>: The numbers $\varepsilon_g(n)$ are determined by the formula

$$\varepsilon_g(n) = \frac{(2n)!}{(n+1)!(n-2g)!} \cdot \text{coefficient of } x^{2g} \text{ in } \left(\frac{x/2}{\tanh x/2}\right)^{n+1}.$$

These three statements can be combined to give Theorem 8.1 as follows. Define a polynomial of degree $d = 3g-1$ in n by

$$F(n) = \frac{(n-1)!}{(n-2g)!} \cdot \text{ coefficient of } x^{2g} \text{ in } \left(\frac{x/2}{\tanh x/2}\right)^{n+1}.$$

Clearly $F(-1) = 0$, so $F(n)/(n+1)$ is still a polynomial in n and can be written

$$F(n) = (n+1) \cdot \sum_{r=1}^{d} \frac{r!}{(2r)!} \kappa_g(r) \cdot (n-1)(n-2)\cdot\ldots\cdot(n-r+1),$$

for some numbers $\kappa_g(r)$. The definition of F and Theorem 8.5 give:

$$\text{(c)} \quad \varepsilon_g(n) = \frac{(2n)!}{n!} \sum_{r=1}^{d} \frac{r!}{(2r)!} \frac{\kappa(r)}{(n-r)!}.$$

the relationship between κ, ε, μ and λ can be understood best by using the generating functions

$$K(x) = \sum_{n\geq 0} \kappa_g(n)x^n, \quad E(x) = \sum_{n\geq 0} \varepsilon_g(n)x^n, \quad M(x) = \sum_{n\geq 0} \mu_g(n)x^n,$$

$$L(x) = \sum_{n\geq 0} \lambda_g(n)x^n.$$

Formulas (a) and (b) of Lemma 8.4 and formula (c) above may then be used to show

$$L(x) = \frac{1}{1+x} M\left(\frac{x}{1+x}\right) = \frac{1}{(1+x)(1+2x)} E\left(\frac{x(1+x)}{(1+2x)^2}\right) = \frac{1}{1+x} K(x(1+x)).$$

Expanding gives

$$\lambda_g(n) = \sum_{r=1}^{d} \kappa_g(r)\binom{r-1}{n-r},$$

and Theorem 8.3 now yields

$$\chi(\Gamma_g^1) = \sum_{n\geq 1} (-1)^{n-1}\lambda_g(n)/2n = \sum_{r=1}^{d} \kappa_g(r) \sum_{n=r}^{2r-1} \frac{(-1)^{n-1}}{2n}\binom{r-1}{n-r}$$

$$= \sum_{r=1}^{d} \kappa_g(r) \frac{(-1)^{r-1}}{2} \frac{(r-1)!^2}{(2r-1)!} = F(0) = -B_{2g}/2g$$

as required.

§2. Counting the Cells of Y

The action of Γ_g^1 on Y is cellular and may be used to compute $\chi(\Gamma_g^1)$. Suppose $\{\sigma_p^i\}$ is a set of representatives for the orbits of the p-cells of Y; thus every p-cell is identified to one of the representatives and no two are identified to one another. Then we have the following formula of Quillen ([S], prop. 11):

$$\chi(\Gamma_g^1) = \sum_p (-1)^p \sum_i \chi(G_p^i)$$

where G_p^i is the stabilizer of the cell σ_p^i.

A p-cell of Y is determined by a rank $n = 6g-3-p$ arc-system which fills F; the dual to this arc system is the graph $\Omega \subset F$ whose complement is a disk centered at $*$. Splitting F along Ω gives a 2n-gon P_n with its center at $*$ and an identification of F with $P_n/\sim$ where $\sim$ is an edge pairing on P_n. The pairing $\sim$ satisfies the conditions to be counted for $\lambda_g(n)$ since Ω has valence ≥ 3 at each vertex. The stabilizer of this p-cell is the group of rotational symmetries of $\sim$; it is finite of order $\frac{2n}{m}$ and its Euler characteristic is $(2n/m)^{-1}$.

The pairings of the edges of P_n occuring in the count for $\lambda_g(n)$ may be partitioned into equivalence classes, two pairings being equivalent if they differ by a rotation of P_n. Choose a representative for each equivalence class, pair the sides of P_n and identify the result with F so that the center of P_n is matched with $*$. This picks out a $p = 6g-3-n$ cell σ_p^i for each class and $\{\sigma_p^i\}$ is a set of representatives for the action of Γ_g^1 on Y. If there are m elements in the equivalence class, the identification will have a cyclic symmetry of order $\frac{2n}{m}$ and the corresponding σ^i will have isotropy group which is cyclic of order $\frac{2n}{m}$. Counting $(\frac{2n}{m})^{-1}$ for each σ^i gives the same answer as counting each of the m elements in each equivalence class with weight $1/2n$. Thus

$$\sum_i \chi(G_p^i) = \lambda_g(n)/2n.$$

Theorem 8.3 follows.

§3. Combinatorics

Now we give the proof of Theorem 8.5.

i) Colorings

A *k-coloring* of the vertices of P_n is a map ϕ from the vertices of P_n to a fixed set of cardinality k, called the set of colors. Define $C(n,k)$ to be the number of pairs (ϕ,τ) where ϕ is a k-coloring of the vertices of P_n and τ is an identification of the edges of P_n which is compatible with ϕ (thus two edges may be identified by τ only if the left end of each has the same color as the right end of the other). If we first identify by τ, the number of inequivalent vertices is $n+1-2g$ where g is the genus of the resulting surface. These vertices can be colored in k^{n+1-2g} ways, so we have shown

$$\text{(d)}\quad C(n,k) = \sum_{g=0}^{n/2} \varepsilon_g(n)k^{n+1-2g}.$$

It is easy to see that the numbers $C(n,k)$ determine the numbers $\varepsilon_g(n)$. In fact, Theorem 8.5 is implied by:

Theorem 8.6: $C(n,k) = (2n-1)!!D(n,k)$

where $D(n,k)$ is defined by the generating functions

$$\text{(e)}\quad 1 + 2\sum_{n=0}^{\infty} D(n,k)x^{n+1} = \left(\frac{1+x}{1-x}\right)^k$$

and $(2n-1)!! = (2n-1)(2n-3)\cdot\ldots\cdot5\cdot3\cdot1$.

Proof of 8.5: To see how 8.6 implies 8.5, differentiate (e) to obtain:

$$(n+1)D(n,k) = k\cdot\operatorname{Res}_{x=0}\left[\frac{1}{x^{n+1}}\left(\frac{1+x}{1-x}\right)^k \frac{dx}{1-x^2}\right].$$

Making the substitution $x = \tanh\frac{t}{2}$ gives

$$(n+1)D(n,k) = \frac{1}{2}\,k\cdot \mathrm{Res}_{t=0}\left[\left(\frac{1}{\tanh t/2}\right)^{n+1} e^{kt}\,dt\right]$$

$$= 2^n k\cdot \text{Coefficient of } t^n \text{ in } e^{kt}\left(\frac{t/2}{\tanh t/2}\right)^{n+1}$$

$$= 2^n k\cdot \sum_{r=0}^{n} \frac{k^r}{r!} \text{ Coefficient of } t^{n-r} \text{ in } \left(\frac{t/2}{\tanh t/2}\right)^{n+1}.$$

Since $\frac{t/2}{\tanh t/2}$ is an even power series, only the coefficients where $n-r$ is even, say $n-r=2g$, are nonzero. Thus the last equality (multiplied by $\frac{(2n-1)!!}{n+1} = \frac{(2n)!}{2^n(n+1)!}$) may be written

$$(2n-1)!!D(n,k) = \frac{(2n)!}{(n+1)!}\sum_{g=0}^{n/2}\frac{k^{n+1-2g}}{(n-2g)!}\times$$

$$\text{Coefficient of } t^{2g} \text{ in } \left(\frac{t/2}{\tanh t/2}\right)^{n+1}. \qquad \square$$

This we are reduced to proving Theorem 8.6.

ii) Full Colorings

A *full* *k-coloring* of the vertices of P_n is a k-coloring ϕ in which every color is used at least once (ϕ is surjective). Let $C_0(n,k)$ be the number of pairs (ϕ,τ) where ϕ is a full k-coloring of the vertices of P_n and τ is a compatible edge pairing. Since the number of inequivalent vertices under any τ is $n+1-2g$ we have

(f) $C_0(n,k) = 0$ if $k > n+1$.

It is easy to relate $C(n,k)$ and $C_0(n,k)$, the formula is

(g) $C(n,k) = \sum_{\ell=0}^{k} \binom{k}{\ell} C_0(n,\ell)$.

We will use formulas (f) and (g) together with the following lemma to prove Theorem 8.6.

Lemma 8.7. The function $D(n,k)$ is a polynomial of degree $k-1$ in n

Proof of 8.6. Formula (g) may be inverted to give

$$C_0(n,k) = \sum_{\ell=0}^{k} (-1)^{k-\ell}\binom{k}{\ell} C(n,\ell).$$

Therefore $C_0(n,k) = (2n-1)!!D_0(n,k)$ where $D_0(n,k) = \sum(-1)^{k-\ell}\binom{k}{\ell}D(n,k$

is again a polynomial of degree $k-1$ in n. Formula (f) says that $D_0(n,k)$ vanishes for $n = 0,1,\dots,k-2$, so $D_0(n,k) = \alpha_k\binom{n}{k-1}$ where α_k is independent of n.

When $n = k-1$, $C_0(k-1,k)$ may be computed directly: the only possible identifications of the sides of a $2(k-1)$ gon which admits a full k-coloring are those giving a surface of genus 0, hence $C_0(k-1,k) = k!\,\varepsilon_0(k-1)$. The recursion

$$\varepsilon_0(n) = \sum_{a+b=n-1} \varepsilon_0(a)\varepsilon_0(b)$$

with initial value $\varepsilon_0(1) = 1$ is easy to see geometrically; it can be solved to give $\varepsilon_0(n) = \binom{2n}{n}/(n+1)$ (the $n^{\underline{th}}$ Catalin number). Working back we find $\alpha_k = 2^{k-1}$, $C_0(n,k) = (2n-1)!!\,2^{k-1}\binom{n}{k-1}$ and so

$$C(n,k) = (2n-1)!! \sum_{\ell\geq 1} 2^{\ell-1}\binom{k}{\ell}\binom{n}{\ell-1}.$$

Expanding $\left(\frac{1+x}{1-x}\right)^k = \left(1+\frac{2x}{1-x}\right)^k$ by the binomial theorem we see that this implies Theorem 8.6. □

iii) An Integral Formula For $C(n,k)$

Finally we come to the heart of the combinatorial argument as we carry out the proof of Lemma 8.7. We begin the proof by reversing our earlier counting procedure; first we color the vertices of P_n and then we make the identifications.

Let ϕ be a coloring of the vertices of P_n and let n_{ij} be the number of edges of P_n which are colored $i-j$ (where i and j are in the color set). Any edge identification τ must identify an edge colored $i-j$ with one colored $j-i$ but the order the edges occur is unimportant. If $n_{ij} \neq n_{ji}$ for some $i \neq j$ or if for some i n_{ii} is odd, then there are no compatible identifications. Thus we want to count the cases where the matrix $N = (n_{ij})$ is even and symmetric, and when it is the number of compatible edge identifications is $\prod_{i<j} n_{ij}!\,\prod_i (n_{ii}-1)!!$. Therefore

$$C(n,k) = \sum_N C(N)\,\varepsilon(N),$$

with $N=(n_{ij})$ where the n_{ij} are non-negative integers such that $\sum n_{ij} = 2n$, $C(N)$ is the number of k-colorings of P_n having n_{ij} edges colored $i-j$ and

$$\varepsilon(N) = \begin{cases} \prod_{1\le i<j\le k} n_{ij}! \times \prod_{i=1}^{k} (n_{ii}-1)!!, & N \text{ even and symmetric,} \\ 0 & , \text{ otherwise.} \end{cases}$$

To understand the number $c(N)$ associate to each coloring of $\mathcal{P}_n$ the product $z_{i_1 i_2} z_{i_2 i_3} \dots z_{i_{2n} i_1}$ where i_j is the color of the $j\underline{th}$ vertex. Allowing the variables to commute we see that this equals $\prod_{i,j} z_{ij}^{n_{ij}}$ which we denote Z^N with $Z = (z_{ij})$, $1 \le i,j \le k$. Working out $\operatorname{tr}(Z^{2n})$ explicitly demonstrates the formula

$$\operatorname{tr}(Z^{2n}) = \sum_N C(N) Z^N .$$

The next step is to introduce an integral representation for the function $\varepsilon(N)$. One uses the fact that

$$\frac{1}{\pi}\int_{-\infty}^{\infty}\int_{-\infty}^{\infty} (x+iy)^n (x-iy)^m e^{-x^2-y^2}\, dxdy = \delta_{nm} n! \quad \text{and}$$

$$\frac{1}{\sqrt{2\pi}}\int_{-\infty}^{\infty} x^n e^{-x^2/2}\, dx = \begin{cases} 0 & n \text{ odd} \\ (n-1)!! & n \text{ even} \end{cases} \quad \text{to see}$$

$$\varepsilon(N) = 2^{-k/2}\pi^{-k^2/2}\int_{H_k} Z^N e^{-\frac{1}{2}\operatorname{tr}(Z^2)}\, d\mu_H$$

where H_k is the space of all $k \times k$ Hermitian matrices $Z = (z_{ij})$ (so $H_k \cong \mathbb{R}^{k^2}$ with variables $x_{ij}\,(i \le j)$ and $y_{ij}\,(i<j)$, and

$$z_{ij} = \begin{cases} x_{ii} & i=j, \\ x_{ij} + \sqrt{-1}y_{ij} & i<j, \\ x_{ij} - \sqrt{-1}y_{ij} & i>j,) \end{cases}$$

and $d\mu_H = \prod_{i\le j} dx_{ij} \prod_{i\le j} dy_{ij}$ is Euclidean volume. Thus we have the integral representation

$$C(n,k) = 2^{-k/2}\pi^{-k^2/2}\int_{H_k} \operatorname{tr}(Z^{2n}) e^{-\frac{1}{2}\operatorname{tr}(Z^2)}\, d\mu_H . \qquad (*)$$

iv) Evaluating the Integral

Let $T_k \subset H_k$ be the diagonal matrices ($k \times k$ with real entries), U_k be the unitary group, Δ_k be the diagonal elements of U_k and W the group of $k \times k$ permutation matrices. Any Hermitian matrix Z is conjugate under U_k to an element of T_k and for all $t \in T_k$ with distinct non-zero entries (therefore for almost all t) the choice of $u \in U_k$ with $Z = u^{-1}tu$ is unique up to left multiplication by an element of $\Delta_k \cdot W$. Define

$$T_k \times \Delta_k \backslash U_k \longrightarrow H_k, \quad \text{by}$$
$$(t,u) \longmapsto u^{-1}tu;$$

this map is generically a covering of degree $k!$

We use the map to make a change of variables in (*) above. It is not difficult to see that

$$d\mu_H = \prod_{i<j} (t_i - t_j)^2 \cdot d\mu_{\Delta\backslash u} \cdot d\mu_T$$

where $d\mu_T$ is Euclidean volume on $T_k \cong \mathbb{R}^k$ and $d\mu_{\Delta\backslash u}$ is the measure induced on $\Delta\backslash u$ by Haar measure. Combining the above with the fact that the function $F(Z) = tr(Z^{2n})e^{-\frac{1}{2}tr(Z^2)}$ is invariant under the action of U_k on H_k by conjugation, we have

$$\int_{H_k} F(Z)\,d\mu_H = \frac{1}{k!}\int_{T_k}\int_{\Delta_k\backslash U_k} F(u^{-1}tu) \prod_{i<j} (t_i-t_j)^2 d\mu_{\Delta\backslash U}\, d\mu_T$$

$$= C_k \int_{-\infty}^{\infty} \cdots \int_{-\infty}^{\infty} F\begin{pmatrix} t_1 & & 0 \\ & \ddots & \\ 0 & & t_k \end{pmatrix} \prod_{1\le i<j\le k} (t_i-t_j)^2 dt_1 \ldots dt_k$$

for some C_k which does not depend on n. This shows

$$C(n,k) = C_k' \int_{\mathbb{R}^k} (t_1^{2n} + \ldots + t_k^{2n}) e^{-\frac{1}{2}(t_1^2 + \ldots t_k^2)} \prod_{1\le i<j\le k} (t_i-t_j)^2 dt_1 \ldots dt_k$$

where C_k' is also independent of n.

By symmetry of the integrand, we may replace $t_1^{2n} + \ldots + t_k^{2n}$ by kt_1^{2n} without changing the value of the integral. Write

$$\prod_{i<j} (t_i - t_j)^2 = \sum_{r=0}^{2k-2} a_r t_1^r$$

where each a_r is a function of $t_2,\ldots,t_k$, and perform the integration over t_1 (using the integral representation for $(n-1)!!$ introduced earlier) to obtain

$$C(n,k) = \sum_{s=0}^{k-1} \alpha_{k,s}(2n+2s-1)!!$$

where $\alpha_{k,r}$ is independent of n. Since $(2n+2s-1)!! = (2n-1)!!$ times a polynomial of degree s in n, we have proven Lemma 8.7 and therefore the main result. □

REFERENCES

[A] M. F. Atiyah, The signature of fibre bundles, Global Analysis, Papers, in Honor of K. Kodaira, Princeton University Press, (1969).

[Ba] W. L. Bailey, On the imbedding of V-manifolds in projective space, Amer. J. Math., 79 (1957), 403-430.

[BiEc] R. Bieri and B. Eckmann, Groups with homological duality generalizing Poincare duality, Inv. Math., 20 (1973), 103-124.

[BLM] J. Birman, A. Lubotsky, J. McCarthy, Abelian and solvable subgroups of the mapping class group, Duke Math. J. 50 (1983), 1107-1120.

[B] A. Borel, Cohomologie de sous-groupes discrets et representations de groupes demisimple, Asterisque, 32-33 (1976), 73-112.

[BS] A. Borel and J. P. Serre, Corners and arithmetic groups, Comm. Math. Helv., 48 (1973), 436-491.

[BE] B. Bowditch and D. Epstein, On Natural Triangulations Associated to a Punctured Surface, preprint (1984).

[Br] K. Brown, Euler characteristics of discrete groups and G-spaces, Inv. Math., 27 (1974), 229-64.

[C] R. Charney, Homology stability for Gl_n of a Dedekind domain, Inv. Math., 56 (1980), 1-17.

[CL] R. Charney and R. Lee, Characteristic classes for the classifying space of Hodge structures, preprint (1984).

[CL1] R. Charney and R. Lee, Cohomology of the Satake compactification, Topology 22 (1983), 389-423.

[D] M. Dehn, Die gruppe der abbildungsklassen, Acta Math., 69 (1971), 81-115.

[EE] C. Earle and J. Eells, The diffeomorphism group of a compact Riemann surface, Bull. A.M.S., 73 (1967), 557-559.

[G] F. P. Gardiner, Schiffer's interior variation and quasiconformal mapping, Duke Math. J., 42 (1975), 371-380.

[GM] H. Glover and G. Mislin, On the stable cohomology of the mapping class group, preprint (1985).

[Gold] W. Goldman, The symplectic nature of fundamental groups of surfaces, Adv. in Math. 54 (1984), 200-225.

[Gr] E. Grossman, On the residual finiteness of certain mapping class groups, preprint.

[Hard] G. Harder, A Gauss-Bonnet formula for discrete arithmetically defined groups, Annales Sci. E.N.S.

[H1] J. Harer, The second homology group of the mapping class group of an orientable surface, Inv. Math., 72 (1982), 221-239.

[H2] J. Harer, Stability of the homology of the mapping class groups of orientable surfaces, Ann. of Math., 121 (1985), 215-249.

[H3] J. Harer, The virtual cohomological dimension of the mapping class group of an orientable surface, Invent. Math. 84 (1986), 157-176.

[HZ] J. Harer and D. Zagier, The Euler characteristic of the moduli space of curves, Invent. Math. 85 (1986), 457-485.

[Ha] J. Harris, Families of smooth curves, preprint, (1984).

[HM] J. Harris and D. Mumford, On the Kodaira dimension of the moduli space of curves, Inv. Math., 67 (1982), 23-86.

[Har] W. Harvey, Boundary structure for the modular group, Riemann Surfaces and Related Topics: Proc. 1978 Stony Brook Conf., Princeton University Press, (1978), 245-251.

[HT] A. Hatcher and W. Thurston, A presentation of the mapping class group of a closed, orientable surface, Topology, 19 (1980), 221-237.

[K] S. Kerckhoff, The Nielson realization problem, Ann. of Math. (2) 117 (1983), 235-265.

[L1] D. Long, A note on the normal subgroups of mapping class groups, preprint, (1985).

[L2] D. Long, Maximal subgroups in mapping class groups, preprint, (1985).

[Mas] H. Masur, The extension of the Weil-Petersson metric to the boundary of Teichmüller space, Duke Math. J., 43 (1976), 623-635.

[McCa] J. McCarthy, Normalizers and centralizers of pseudo-Anosov mapping classes, preprint, (1984).

[Mc] J. McCool, Some finitely presented subgroups of the automorphism group of a free group, J. Alg, 35 (1975), 205-213.

[Mi] E. Miller, The homology of the mapping class group, J. Diff. Geom. 24 (1986), 1-14.

[MM] J. Milnor and J. Moore, On the structure of Hopf algebras, Ann. of Math., 81 (1965), 211-264.

[Mor] S. Morita, Characteristic classes of surface bundles I, preprint, (1984).

[Mos] L. Mosher, Pseudo-Anosovs on punctured surfaces, thesis, Princeton University, (1983).

[M] D. Mumford, Abelian quotients of the Teichmüller modular group, J. d'Anal. Math., 18 (1967), 227-244.

[M1] D. Mumford, Towards an enumerative geometry of the moduli space of curves, Arithmetic and Geometry, Vol. II, Birkhäuser (1983), 271-328.

[P] J. Powell, Two theorems on the mapping class group of a surface, Proc. AMS, 68 (1978), 347-350.

[Q] D. Quillen, M.I.T. lectures, (1974-5).

[R] H. Royden, Automorphisms and isometries in Teichmüller space, in Ann. of Math. Studies 66, (1971).

[Se] J. P. Serre, Cohomologie des groupes discrets, Prospects in Mathematics, Ann. of Math. Studies no. 70, Princeton University Press (1971), 77-169.

[Se1] J. P. Serre, Arithmetic groups, Homological Group Theory, edited by C.T.C. Wall, Camb. Univ. Press, (1979).

[S] K. Strebel, On quadratic differentials with closed trajectories and second order poles, J. d'Anal. Math., 19 (1967), 373-382.

[S1] K. Strebel, Quadratic Differentials, Erg. der Math. und ihrer Grenz. 5, Springer-Verlag (1984).

[T1] W. Thurston, Geometry and topology of 3-manifolds, Princeton lectures, (1978).

[T2] W. Thurston, A spine for Teichmüller space, preprint, (1984).

[T3] W. Thurston, On the geometry and dynamics of diffeomorphisms of surfaces, preprint.

[V] K. Vogtmann, Spherical posets and homology stability for $O_{n,n}$, Topology, 20 (1981), 119-132.

[W] J. Wagoner, Stability for the homology of the general linear group of a local ring, Topology, 15 (1976), 417-423.

[Wa] B. Wajnryb, A simple presentation for the mapping class group of an orientable surface, Israel J. Math. 45 (1983), 157-174.

[Wol1] S. Wolpert, The Fenchel-Neilsen deformation, Ann. of Math., 115 (1982), 501-528.

[Wol2] S. Wolpert, A projective embedding of the moduli space of stable curves, preprint (1983).

[Wol3] S. Wolpert, On the symplectic geometry of deformations of a hyperbolic surface, Ann. of Math., 117 (1983), 207-234.

[Wol4] S. Wolpert, Thurston's Riemannian metric for Teichmüller space, J. Diff. Geom. 23 (1986), 143-174.

[Wol5] S. Wolpert, On the Weil-Petersson geometry of the moduli space of curves, Amer. J. Math., 107 (1985), 969-998.

FONDAZIONE C.I.M.E.
CENTRO INTERNAZIONALE MATEMATICO ESTIVO
INTERNATIONAL MATHEMATICAL SUMMER CENTER

"Inverse Problems"

is the subject of the First 1986 C.I.M.E. Session.

The Session, sponsored by the Consiglio Nazionale delle Ricerche and the Ministero della Pubblica Istruzione, will take place under the scientific direction of Prof. GIORGIO TALENTI (Università di Firenze) at Villa «La Querceta», Montecatini Terme (Pistoia), Italy, ***from May 28 to June 5, 1986.***

Courses

a) ***Inverse Eigenvalue Problems.*** (6 lectures in English).
Prof. Victor BARCILON (University of Chicago).

— Introduction. Eigenvalue problems arising in the study of vibrating systems: string, beam, elastic earth. Impulse response; equivalent data sets. Asymptotic behaviour of eigensolutions.
— Oscillatory kernels and Chebyshev systems. The work of Kellogg, M.G. Krein, Karlin.
— Inverse Sturm-Liouville problem. Borg uniqueness theorem. Improvements of Levinson and Marchenko. End-point formules.
— Inverse problem for the vibrating string. Discretization and construction algorithms. Existence results. End-point formulas and reduction to quadratures.
— Inverse problem for the vibrating beam. Discretization: oscillatory matrices.
— Inverse problem for the vibrating beam. Existence theorem.

References

- BARCILON, V., Inverse problem for the vibrating beam in the free-clamped configuration, Phil. Trans. Roy Soc. London, A 304 (1982), 211-251.
- BARCILON, V., Explicit solution of the inverse problem for a vibrating string, J. Math. Anal. Appl., 93 (1983), 222-234.
- BORG, G., Eine Umkehrung der Sturm-Liouvilleschen Eigenwertaufgabe, Acta Math., 78 (1946), 1-96.
- GANTMAKHER, F.P., The Theory of Matrices. Chelsea, New York, 1959.
- GANTMAKHER, F.P., and KREIN, M.G., Oscillation Matrices and Kernels and Small Vibrations of Mechanical Systems. Office of Technical Services, Washington, D.C., 1961.
- GELFAND, I.M., and LEVITAN, B.M., On a simple identity for the eigenvalues of a differential operator of the second order, Dokl. Akad. Nauk SSSR, 88 (1953), 593-596.
- GLADWELL, G.M. L., The inverse problem for the vibrating beam, Proc. Roy. Soc. London, A 393 (1984), 277-295.
- KARLIN, S., Total positivity. Stanford Univ. Press, 1968
- KELLOGG, O.D., The oscillations of functions of an orthogonal set, Amer. Math. Soc., 38 (1916), 1-5.
- KELLOGG, O.D., Orthogonal function sets arising from integral equations, Amer. Math. Soc., 40 (1918), 145-154.
- KREIN, M.G., Sur les vibrations propres des tiges dont l'une des extremités est encastrée et l'autre libre, Comm. Soc. Math. Kharkoff, 12 (1935), 3-11
- KREIN, M.G., On inverse problems for an inhomogeneous string, Dokl. Akad. Nauk SSSR, 82 (1952), 669-672 (in Russian).
- KREIN, M.G., Some new problems in the theory of Sturmian systems, Prik. Mat. Mekh., 16 (1952), 555-568 (in Russian).
- LEVINSON, N., The inverse Sturm-Liouville problem, Mat. Tidsskr. B (1949), 25-30.
- MARCHENKO, V.A., Some questions in the theory of one-dimensional linear differential operators of the second order, Amer. Math. Soc. Trans., 101 (1973), 1- 104.

b) ***Regularization Methods for Linear Inverse Problems.*** (6 lectures in English).
Prof. Mario BERTERO (Università di Genova).

Contents: Linear inverse problems and linear operator equations. Examples of ill- posed inverse problems. Generalized inverses of linear operators. Approximate solutions satisfying prescribed constraints. Ivanov constrained pseudosolutions and Miller least squares method. Regularizing algorithms: general definition and examples (Tikhonov regularizer, spectral windows etc.). Choice of the regularization parameter and the discrepancy principle. Stability estimates. Linear inverse problems with discrete data. Applications to first kind Fredholm integral equations.

Basic literature:

- GROETSCH, C.W., Genralized Inverses of Linear Operators. Dekker, New York, 1977
- GROETSCH, C.W., The Theory of Tikhonov Regularization for Fredholm Equations of the First kind Pitman, Boston, 1984.
- LAVRENTIEV, M.M., Some Improperly Posed Problems of Mathematical Physics. Springer, Berlin, 1967.
- MOROZOV, V.A., Methods for Solving Incorrectly Posed Problems Springer, Berlin, 1984.
- NASHED, M.Z. (ed.), Generalized Inverses and Applications. Academic Press, New York, 1976.
- TIKHONOV, A.N. and ARSENIN, V.Y., Solutions of Ill-Posed Problems Winson/ Wiley, Washington, 1977.

c) ***Tomography*** (6 lectures in English).
Prof. Alberto F. GRUNBAUM (Universiy of California, Berkeley).

d) ***Numerical Treatment of Ill-Posed Problems.*** (6 lectures in English).
Prof. Frank NATTERER (Univ. Münster).

Ill-posed problems are problems whose solutions are not well defined in the sense that they do not exist at all, or that they are not unique, or that they do not depend continuously on the data.

The session starts with a brief introduction into the theory of ill-posed problems, the practical difficulties, and typical areas where such problems arise.

We then present basic regularization techniques, such as truncated singular value decomposition, Tikhonov regularization, iteration, coarse discretization, and (constrained) optimization. The numerical implementation of the first two of these techniques is discussed in detail. A crucial point in both techniques is the determination of a "good" parameter controlling the trade-off between a-priori-knowledge and information contained in the data.

At the end of the session we present numerical results for selected problems in tomography.

References

- TIKHONOV, A.N. - ARSENIN, V.Y., Solution of Ill-Posed Problems. Wiley 1977.
- RAMM, A., Theory and application of some new classes of integral equations Springer 1980
- BERTERO, M., Problemi lineari non ben posti e metodi di regolarizzazione Consiglio Nazionale delle Ricerche, Firenze 1982.
- LAVRENTEV, M.M., ROMANOV, V.G., SISATSKIJ, S.P., Problemi non ben posti in Fisica matematica e Analisi, Consiglio Nazionale delle Ricerche, Firenze 1983
- GROETSCH, C.W., The theory of Tikhonov regularization for Fredholm equation of the first kind. Pitman 1984.

FONDAZIONE C.I.M.E.
CENTRO INTERNAZIONALE MATEMATICO ESTIVO
INTERNATIONAL MATHEMATICAL SUMMER CENTER

"Mathematical Economics"

is the subject of the Second 1986 C.I.M.E. Session.

The Session, sponsored by the Consiglio Nazionale delle Ricerche and the Ministero della Pubblica Istruzione, will take place under the scientific direction of Proff. ANTONIO AMBROSETTI (Università di Venezia), FRANCO GORI (Università di Firenze), ROBERTO LUCCHETTI (Università di Milano) at Villa «La Querceta», Montecatini Terme (Pistoia), Italy, ***from June 25 to July 3, 1986.***

Courses

a) ***Variational Problems arising from Mathematical Economics.*** (6 lectures in English).
Prof. Ivar EKELAND (CEREMADE, Univ. Paris IX).

The lectures will focus on two topics: 1. Infinite horizon optimization problems arising from Ramsey models, 2. Generalization of Hamilton-Jacobi theory arising from the theory of incentive.

Prerequisites: 1. Some familiarity with convex optimization is advisable. 2. I. Ekeland-R. Temam, Analyse convexe et problèmes variationnels, Dunod 1974.

b) ***Differentiability Techniques in the Theory of General Economic Equilibrium.*** (6 lectures in English).
Prof. Andreu MAS-COLELL (Harvard University).

Outline of content:

Applications of Calculus and Differential Topology methods to the problem of existence, optimality and local uniqueness of price equilibria will be reviewed. Comparative statics will also be considered. Both static and sequential models of economic equilibrium will be discussed

Basic references

- E. DIERKER, Topological Methods in Walrasian Economics, Lecture Notes in Economics and Mathematical Sciences, 92, 1974, Springer Verlag.
- S. SMALE, Global Analysis and Economics Ch. 8 in "Handbook of Mathematical Economics", vol. II, 1981, ed. K. Arrow and M. Intriligator, pp. 331-370. North-Holland
- A. MAS-COLELL, The theory of general economic equilibrium: A differentiable approach. Cambridge University Press, 1985.

Prerequisites

The economics will be selfcontained althought it would be helpful to have some familiarity with a graduate textbook such as: H. Varian, Microeconomics (Norton) or E. Malinvaud, Microeconomic theory (North-Holland).

For the mathematics, advanced Calculus is all that will be really needed. However, it would be convenient if the basics of real analysis could be taken for granted and, also, if students had become familiar with the simple facts of transversality theory (as gotten for example from: J. Milnor, Topology from the Differentiable View-point, 1965, University Press of Virginia, or the first half of: Guillemin, V. and A. Pollak, Differential Topology, NJ, Prentice Hall).

c) ***Dynamical General Equilibrium Models.*** (6 lectures in English).
Prof. Jose A. SCHEINKMAN (University of Chicago).

This course will deal with dynamic general equilibrium economic models, i.e., the integration of general equilibrium analysis and dynamics. Special attention will be placed on the implications of these models to the phenomena of macroeconomic fluctuations. In order to avoid excessive prerequisites we will start with a short survey of the relevant aspects of general equilibrium theory and the theory of economic growth.

Contents

— Arrow-Debreu Theory.
— Models of Economic Growth under Uncertainty as Dynamic Equilibrium Models.
— Incomplete Market Models and Economic Fluctuations.

Main References

- ARROW, K., "The Role of Securities in the Optimal Allocation of Risk Bearing", in "Essay in the Theory of Risk-Bearing", Markham Press.
- BROCK, W.A. and MIRMAN, L.J., Optimal Economic Growth and Uncertainty: The Discounted Case, Journal of Economic Theory, 4 (1972), 233-240.
- LUCAS, ROBERT E., Asset Prices in an Exchange Economy, Econometrica, 46 (1978), 1429-1445.
- LUCAS, ROBERT E., Models of Business Cycles, (mimeo) 1985.
- SAMUELSON, P., An exact Consumption-Loan Model of Interest ..., Journal of Political Economy, 66 (1958), 67-82.
- SCHEINKMAN, JOSE, General Equilibrium Models of Economic Fluctuations: A Survey Theory, (mimeo) 1984.
- SCHEINKMAN, J. and WEISS, L., Borrowing Constraints and Aggregate Economic Activity, Econometrica, January 1986.

d) ***Topics in Noncooperative Game Theory.*** (6 lectures in English)
Prof. Shmuel ZAMIR (Hebrew Univ. Jerusalem).

— Basic concepts and Results: Games in Extensive form; Games in Strategic form; Nash Equilibria; Zero-Sum Games - Minimax Theorems.
— Multistage Games: Repeated Games; Stochastic Games; Blackwell's Approachability Nation.
— Incomplete Information: The Universal Beliefs Space; Common Knowledge; Consistency of beliefs.
— Zero-Sum Repeated Games with Incomplete Information: Minimax Maxmin and Asymptotic Value; Speed of Convergence, the Variation of a Martingale and the Normal Distribution; Games without Recursive Structure.

Basic References

- OWEN, G., 'Game Theory', Second Edition, Academic Press, 1982.
- BURGER, E., 'Introduction to the Theory of Games', Prentice-Hall, 1963.
- VON NEUMAN, L. and O. MORGENSTERN, 'Theory of Games and Economic Behavior', Princeton University Press, 1944, 1947.
- PARTHASARATHY, T. and T.E.S. RAGAVAN, 'Some Topics in Two-Person Games', American Elsevier, 1971.
- SORIN, S., 'An introduction to Two-Person Zero-Sum Repeated Games with Incomplete Information', Technical Report No. 312, 1980, Institute for Mathematical Studies in the Social Sciences, Stanford University.
- MERTENS, J.F. and S. ZAMIR, 'Formulation of Bayesian Analysis for Games with Incomplete Information', International Journal of Game Theory 14 (1985), 1-29.

FONDAZIONE C.I.M.E.
CENTRO INTERNAZIONALE MATEMATICO ESTIVO
INTERNATIONAL MATHEMATICAL SUMMER CENTER

"Combinatorial Optimization"

is the subject of the Third 1986 C.I.M.E. Session.

The Session, sponsored by the Consiglio Nazionale delle Ricerche and the Ministero della Pubblica Istruzione, will take place under the scientific direction of Prof. BRUNO SIMEONE (Università di Roma "La Sapienza") at the Centro di Cultura Scientifica A. Volta, «Villa Olmo», Como (Italy), ***from August 25 to September 2, 1986.***

Program outline

The CIME Summer School on "Combinatorial Optimization" aims to present recent results and current trends in this area.

Combinatorial Optimization has a peculiar location in the map of Applied Mathematics, being placed in an interzone in the middle of Combinatorics, Computer Science and Operations Research. From a mathematical point of view, it draws on pure combinatorics, including graphs and matroids, on Boolean algebras and switching functions, partially ordered sets, group theory, linear algebra convex geometry and probability theory, as well as other tools.

Over the past years, a substantial amount of research has been devoted to the connections between Combinatorial Optimization and theoretical Computer Science, and in particular to computational complexity and algorithmic issues. Quite often, the study of combinatorial optimization problems is motivated by real-life applications, such as scheduling, assignment, location, distribution, routing, districting, design and other Operations Research applications.

Although references to actual applications will be frequently given, the emphasis of the School is on theoretical aspects of Combinatorial Optimization. Special attention will be devoted to polyhedral combinatorics and its connections with combinatorial duality theories and min-max identities; to the study of important classes of functions (either real- or binary-valued) defined on the binary n-cube, and of their significance in 0-1 optimization; to the role of submodularity (a discrete analogue of convexity) in Combinatorial Optimization; to the deep link between greedy algorithms and finite geometries such as matroids and greedoids. As a matter of fact, the interplay between structural and algorithmic properties is one of the main themes addressed by the School.

Courses

a) ***Truth Functions and Set Functions.*** (6 lectures in English).
Prof. Peter L. HAMMER (Rutgers University, USA).

Truth functions and Boolean expressions. Set functions and pseudo-Boolean expressions. Quadratic Boolean expressions. Consistency of quadratic Boolean equations and the Konig-Egervary property of graphs. The parametric form of the general solution of a quadratic Boolean equation.

Quadratic pseudo-Boolean expressions. Maximization of pseudo-Boolean expressions; reduction to the quadratic case. The conflict graphs of a set function. Quadratic graphs and the problem of their recognition.

Linear majorants of quadratic pseudo-Boolean expressions. Roofs and the height of a quadratic function. Optimal roofs and the complement of a quadratic function. Equivalence of quadratic optimization and of the weighted stable set problem in a graph.

Roof duality. Strong and weak persistence.

The gap between the height and the maximum of a quadratic function. Supermodular functions are gap-free. Gap-free functions and the weighted Konig-Egervary property of ghaphs.

Open problems.

b) ***Binary Group Polyhedra, Binary Matroids, and the Chinese Postman Problem.*** (6 Lectures in English).
Prof. Ellis L. JOHNSON (IBM Thomas J. Watson Research Center, USA).

Abstract

The area to be covered has been worked on from different directions by several people. Whitney introduced matroids and defined binary matroids and graphic matroids.

Tutte characterized graphic matroids. Lehman carried matroid idea into the Shannon switching game and did fundamental work on clutters or ports of matroids. He also developed a framework of various equivalent properties for pairs of dual clutters.

Fulkerson showed that there are then pairs of polyhedra with and elegant and simple duality between their facets and vertices. Independently, Gomory developed a theory of group problems for integer programming with some special results for binary group problems. The polyhedron for a special case, the Chinese postman problem, was characterized by Edmonds and Johnson and seen to be an instance of this general framework.

Seymour further characterized binary clutters and those for which a strong integrality property, the max-flow min-cut property, holds for binary clutters. The emphasis here will be on polyhedra, optimization problems, and the binary group approach.

A dual, or blocking, framework is developed in which relations between various results can be seen.

References

- A. LEHMAN, A Solution of the Shannon Switching Game, SIAM J. App Math. 12 (1984), 687-725.
- A. LEHMAN, On the Width-Lenght Inequality, Math. Prog. 17 (1979), 403-417.
- R.E. GOMORY, Some Polyhedra Related to Combinatorial Optimization, Lin Algebra and Applications 2 (1969), 451-558.
- D.R. FULKERSON, Blocking Polyhedra, in 'Graph Theory and Its Applicantions', B. Harris (ed.), Academic Press, N.Y., 1970, 93-112.
- J. EDMONDS and E.L. JOHNSON, Matchings, Euler Tours, and the Chinese Postman Problem, Math. Prog. 5 (1973), 88-124.
- P.D. SEYMOUR, Matroids with the Max-flow Min-cut Property, J. of Comb. Theory B 23 (1977), 189-222.
- G GASTOU and E.L. JOHNSON, Binary Group and Chinese Postman Polyhedra, to appear in Math. Prog..

c) ***Algorithmic Principles and Combinatorial Structures.*** (6 lectures in English).
Prof. Bernhard KORTE (Universität Bonn).

Abstract

This series of lectures discusses several algorithmic principles in combinatorial optimization.

One goal of this survey will be the study of the relations between algorithmic approaches and the underlying combinatorial structure. A combinatorial optimization problem can be defined either on a simple structure, but then it needs complicated and sophisticated algorithmic tools. Or it can be defined on a combinatorial structure which is mathematically substantial. Then it needs an easy algorithm (e.g. greedy approach). These trade-offs between structure and algorithms are studied in the first part of the lectures.

The second part is basically related to the greedy algorithm and its variants.

We discuss greedy approaches from an algorithmic as well as from a structural point of view.

Finally, we study greedoids which were introduced as relaxations of matroids by L. Lovasz and the lecturer. It turns out that many "real world" combinatorial optimization problems and structures can be formulated in the framework of greedoids.

We discuss polyhedral, structural and algorithmic aspects of greedoids. Very many special greedoids will be investigated, among which antimatroid have a considerably rich structure.

References

a) General introduction into the subject

- WELSH, D.J.A., Matroid Theory, Academic Press, London, 1976.
- LAWLER, E.L., Combinatorial Optimization: Networks and Matroids, Holt, Reinhard & Wiston, New York, 1976.
- PAPADIMITRIOU, C.H. and STEILITZ, K., Combinatorial Optimization: Algorithms and Complexity, Prentice-Hall, Englewood Cliffs, 1982.

b) Special literature

- KORTE, B. and LOVASZ, L., Polymatroid greedoids. J. Comb. Theory (series B) 38 (1985), 41-72.
- KORTE, B. and LOVASZ, L., Greedoids and linear objections functions. SIAM J. Algebraic Discrete Methods 5 (1984), 229-238.
- KORTE, B. and LOVASZ, L., Shelling structures, convexity and a happy end. in: 'Graph Theory and Combinatorics', Proceedings of the Cambridge Combinatorial Conference in Honour of Paul Erdős, Academic Press, London, 1984, pp. 219-232.
- KORTE, B. and LOVASZ, L., Structural properties of greedoids, Combinatorica 3 (1983), 359-374.
- KORTE, B. and LOVASZ, L., Greedoids - a structural framework for the greedy algorithm. in W.R. Pulleyblank (ed.): Progress in Combinatorial Optimization Proceedings of the Silver Jubilee Conference on Combinatorics. Waterloo, June 1982, Academic Press, London, pp. 221-243

d) ***Submodular Functions, Graphs and Optimization.*** (6 lectures in English).
Prof. Eugene L. LAWLER (University of California, Berkeley).

— Matroids and submodular functions.
— Classical Network flow theory: a review.
— Duality theorems via the Max-flow Min-cut Theorem of polymatroidal network flows.
— The matroid parity problem.
— Linear time algorithms for optimal subgraph problems.
— Solution of combinatorial optimization problems by distributed computation.
— Recent results in scheduling theory.

References

- E.L. LAWLER, Combinatorial Optimization: Networks and Matroids, Holt Reinhart and Wiston, 1976.
- E.L. LAWLER and C.U. MARTEL, Flow Network Formulations od Polymatroidal Optimization Problems, Annals Discrete Math., 16 (1982), 189-200.
- E.L. LAWLER and C.U. MARTEL, Computing Maximal 'Polymatroidal' Network Flows, Math. of Operations Research, 7 (1982), 334-347.
- E.L. LAWLER, A Fully Polynomial Approximation Scheme for the Total Tardiness Problem, Operations Research Letters, 1 (1982), 207-208.
- E.L. LAWLER, Preemptive Scheduling of a Single Machine to Minimize the Sum of Completion Times, to appear in Operations Research Letters.
- E.L. LAWLER, Recent Results in Machine Scheduling Theory, Mathematical programming: The State of the Art, A. Bachem et al., Springer Verlag, 1983, pp. 202-234.
- E.L. LAWLER, P. TONG and V. VAZIRANI, Solving the Weighted Parity problem for Gammoids by Reduction to Graphic Matching, Progress in Combinatorial Optimization, W.R Pulleyblank, ed., Academic Press, 1984, pp. 363-374.
- E.L. LAWLER, Submodular Functions and Polymatroid Optimization, in 'Combinatorial Opimization: Annotated Bibliographies', M. O'hEigerartaigh, J.K. Lenstra, A.H.G. Rinnooy Kan, Eds., J. Wiley, 1985.
- E.L. LAWLER, Why Certain Subgraph Computations Require Only Linear Time, Proc. 26th Annual IEEE Symposium on Foundations of Computer Science, October 1985.

LIST OF C.I.M.E. SEMINARS

		Publisher
1954 -	1. Analisi funzionale	C.I.M.E.
	2. Quadratura delle superficie e questioni connesse	"
	3. Equazioni differenziali non lineari	"
1955 -	4. Teorema di Riemann-Roch e questioni connesse	"
	5. Teoria dei numeri	"
	6. Topologia	"
	7. Teorie non linearizzate in elasticità, idrodinamica, aerodinamica	"
	8. Geometria proiettivo-differenziale	"
1956 -	9. Equazioni alle derivate parziali a caratteristiche reali	"
	10. Propagazione delle onde elettromagnetiche	"
	11. Teoria della funzioni di più variabili complesse e delle funzioni automorfe	"
1957 -	12. Geometria aritmetica e algebrica (2 vol.)	"
	13. Integrali singolari e questioni connesse	"
	14. Teoria della turbolenza (2 vol.)	"
1958 -	15. Vedute e problemi attuali in relatività generale	"
	16. Problemi di geometria differenziale in grande	"
	17. Il principio di minimo e le sue applicazioni alle equazioni funzionali	"
1959 -	18. Induzione e statistica	"
	19. Teoria algebrica dei meccanismi automatici (2 vol.)	"
	20. Gruppi, anelli di Lie e teoria della coomologia	"
1960 -	21. Sistemi dinamici e teoremi ergodici	"
	22. Forme differenziali e loro integrali	"
1961 -	23. Geometria del calcolo delle variazioni (2 vol.)	"
	24. Teoria delle distribuzioni	"
	25. Onde superficiali	"
1962 -	26. Topologia differenziale	"
	27. Autovalori e autosoluzioni	"
	28. Magnetofluidodinamica	"
1963 -	29. Equazioni differenziali astratte	"
	30. Funzioni e varietà complesse	"
	31. Proprietà di media e teoremi di confronto in Fisica Matematica	"

1964 - 32. Relatività generale C.I.M.E.
33. Dinamica dei gas rarefatti "
34. Alcune questioni di analisi numerica "
35. Equazioni differenziali non lineari "

1965 - 36. Non-linear continuum theories "
37. Some aspects of ring theory "
38. Mathematical optimization in economics "

1966 - 39. Calculus of variations Ed. Cremonese, Firenze
40. Economia matematica "
41. Classi caratteristiche e questioni connesse "
42. Some aspects of diffusion theory "

1967 - 43. Modern questions of celestial mechanics "
44. Numerical analysis of partial differential equations "
45. Geometry of homogeneous bounded domains "

1968 - 46. Controllability and observability "
47. Pseudo-differential operators "
48. Aspects of mathematical logic "

1969 - 49. Potential theory "
50. Non-linear continuum theories in mechanics and physics and their applications "
51. Questions of algebraic varieties "

1970 - 52. Relativistic fluid dynamics "
53. Theory of group representations and Fourier analysis "
54. Functional equations and inequalities "
55. Problems in non-linear analysis "

1971 - 56. Stereodynamics "
57. Constructive aspects of functional analysis (2 vol.) "
58. Categories and commutative algebra "

1972 - 59. Non-linear mechanics "
60. Finite geometric structures and their applications "
61. Geometric measure theory and minimal surfaces "

1973 - 62. Complex analysis "
63. New variational techniques in mathematical physics "
64. Spectral analysis "

1974 -	65. Stability problems		Ed. Cremonese, Firenze
	66. Singularities of analytic spaces		"
	67. Eigenvalues of non linear problems		"
1975 -	68. Theoretical computer sciences		"
	69. Model theory and applications		"
	70. Differential operators and manifolds		"
1976 -	71. Statistical Mechanics		Ed. Liguori, Napoli
	72. Hyperbolicity		"
	73. Differential topology		"
1977 -	74. Materials with memory		"
	75. Pseudodifferential operators with applications		"
	76. Algebraic surfaces		"
1978 -	77. Stochastic differential equations		"
	78. Dynamical systems		Ed. Liguori, Napoli and Birkhäuser Verlag
1979 -	79. Recursion theory and computational complexity		Ed. Liguori, Napoli
	80. Mathematics of biology		"
1980 -	81. Wave propagation		"
	82. Harmonic analysis and group representations		"
	83. Matroid theory and its applications		"
1981 -	84. Kinetic Theories and the Boltzmann Equation	(LNM 1048)	Springer-Verlag
	85. Algebraic Threefolds	(LNM 947)	"
	86. Nonlinear Filtering and Stochastic Control	(LNM 972)	"
1982 -	87. Invariant Theory	(LNM 996)	"
	88. Thermodynamics and Constitutive Equations	(LN Physics 228)	"
	89. Fluid Dynamics	(LNM 1047)	"
1983 -	90. Complete Intersections	(LNM 1092)	"
	91. Bifurcation Theory and Applications	(LNM 1057)	"
	92. Numerical Methods in Fluid Dynamics	(LNM 1127)	"
1984	93. Harmonic Mappings and Minimal Immersions	(LNM 1161)	"
	94. Schrödinger Operators	(LNM 1159)	"
	95. Buildings and the Geometry of Diagrams	(LNM 1181)	"
1985 -	96. Probability and Analysis	(LNM 1206)	"
	97. Some Problems in Nonlinear Diffusion	(LNM 1224)	"
	98. Theory of Moduli	(LNM 1337)	"

Note: Volumes 1 to 38 are out of print. A few copies of volumes 23,28,31,32,33,34,36,38 are available on request from C.I.M.E.

1986 -	99. Inverse Problems	(LNM 1225)	Springer-Verlag
	100. Mathematical Economics	(LNM 1330)	"
	101. Combinatorial Optimization	to appear	"
1987 -	102. Relativistic Fluid Dynamics	to appear	"
	103. Topics in Calculus of Variations	to appear	"

LECTURE NOTES IN MATHEMATICS

Edited by A. Dold and B. Eckmann

Some general remarks on the publication of monographs and seminars

In what follows all references to monographs, are applicable also to multiauthorship volumes such as seminar notes.

§1. Lecture Notes aim to report new developments - quickly, informally, and at a high level. Monograph manuscripts should be reasonably self-contained and rounded off. Thus they may, and often will, present not only results of the author but also related work by other people. Furthermore, the manuscripts should provide sufficient motivation, examples and applications. This clearly distinguishes Lecture Notes manuscripts from journal articles which normally are very concise. Articles intended for a journal but too long to be accepted by most journals, usually do not have this "lecture notes" character. For similar reasons it is unusual for Ph.D. theses to be accepted for the Lecture Notes series.

Experience has shown that English language manuscripts achieve a much wider distribution.

§2. Manuscripts or plans for Lecture Notes volumes should be submitted either to one of the series editors or to Springer-Verlag, Heidelberg. These proposals are then refereed. A final decision concerning publication can only be made on the basis of the complete manuscripts, but a preliminary decision can usually be based on partial information: a fairly detailed outline describing the planned contents of each chapter, and an indication of the estimated length, a bibliography, and one or two sample chapters - or a first draft of the manuscript. The editors will try to make the preliminary decision as definite as they can on the basis of the available information.

§3. Lecture Notes are printed by photo-offset from typed copy delivered in camera-ready form by the authors. Springer-Verlag provides technical instructions for the preparation of manuscripts, and will also, on request, supply special staionery on which the prescribed typing area is outlined. Careful preparation of the manuscripts will help keep production time short and ensure satisfactory appearance of the finished book. Running titles are not required; if however they are considered necessary, they should be uniform in appearance. We generally advise authors not to start having their final manuscripts specially tpyed beforehand. For professionally typed manuscripts, prepared on the special stationery according to our instructions, Springer-Verlag will, if necessary, contribute towards the typing costs at a fixed rate.

The actual production of a Lecture Notes volume takes 6-8 weeks.

.../...

§4. Final manuscripts should contain at least 100 pages of mathematical text and should include

- a table of contents
- an informative introduction, perhaps with some historical remarks. It should be accessible to a reader not particularly familiar with the topic treated.
- a subject index; this is almost always genuinely helpful for the reader.

§5. Authors receive a total of 50 free copies of their volume, but no royalties. They are entitled to purchase further copies of their book for their personal use at a discount of 33.3 %, other Springer mathematics books at a discount of 20 % directly from Springer-Verlag.

Commitment to publish is made by letter of intent rather than by signing a formal contract. Springer-Verlag secures the copyright for each volume.

Vol. 1173: H. Delfs, M. Knebusch, Locally Semialgebraic Spaces. XVI, 329 pages. 1985.

Vol. 1174: Categories in Continuum Physics, Buffalo 1982. Seminar. Edited by F. W. Lawvere and S. H. Schanuel. V, 126 pages. 1986.

Vol. 1175: K. Mathiak, Valuations of Skew Fields and Projective Hjelmslev Spaces. VII, 116 pages. 1986.

Vol. 1176: R.R. Bruner, J.P. May, J.E. McClure, M. Steinberger, H_∞ Ring Spectra and their Applications. VII, 388 pages. 1986.

Vol. 1177: Representation Theory I. Finite Dimensional Algebras. Proceedings, 1984. Edited by V. Dlab, P. Gabriel and G. Michler. XV, 340 pages. 1986.

Vol. 1178: Representation Theory II. Groups and Orders. Proceedings, 1984. Edited by V. Dlab, P. Gabriel and G. Michler. XV, 370 pages. 1986.

Vol. 1179: Shi J.-Y. The Kazhdan-Lusztig Cells in Certain Affine Weyl Groups. X, 307 pages. 1986.

Vol. 1180: R. Carmona, H. Kesten, J.B. Walsh, École d'Été de Probabilités de Saint-Flour XIV – 1984. Édité par P.L. Hennequin. X, 438 pages. 1986.

Vol. 1181: Buildings and the Geometry of Diagrams, Como 1984. Seminar. Edited by L. Rosati. VII, 277 pages. 1986.

Vol. 1182: S. Shelah, Around Classification Theory of Models. VII, 279 pages. 1986.

Vol. 1183. Algebra, Algebraic Topology and their Interactions. Proceedings, 1983. Edited by J.-E. Roos. XI, 396 pages. 1986.

Vol. 1184: W. Arendt, A. Grabosch, G. Greiner, U. Groh, H.P. Lotz, U. Moustakas, R. Nagel, F. Neubrander, U. Schlotterbeck, One-parameter Semigroups of Positive Operators. Edited by R. Nagel. X, 460 pages. 1986.

Vol. 1185: Group Theory, Beijing 1984. Proceedings. Edited by Tuan H.F. V, 403 pages. 1986.

Vol. 1186: Lyapunov Exponents. Proceedings, 1984. Edited by L. Arnold and V. Wihstutz. VI, 374 pages. 1986.

Vol. 1187: Y. Diers, Categories of Boolean Sheaves of Simple Algebras. VI, 168 pages. 1986.

Vol. 1188: Fonctions de Plusieurs Variables Complexes V. Séminaire, 1979–85. Edité par François Norguet. VI, 306 pages. 1986.

Vol. 1189: J. Lukeš, J. Malý, L. Zajíček, Fine Topology Methods in Real Analysis and Potential Theory. X, 472 pages. 1986.

Vol. 1190: Optimization and Related Fields. Proceedings, 1984. Edited by R. Conti, E. De Giorgi and F. Giannessi. VIII, 419 pages. 1986.

Vol. 1191: A.R. Its, V.Yu. Novokshenov, The Isomonodromic Deformation Method in the Theory of Painlevé Equations. IV, 313 pages. 1986.

Vol. 1192: Equadiff 6. Proceedings, 1985. Edited by J. Vosmansky and M. Zlámal. XXIII, 404 pages. 1986.

Vol. 1193: Geometrical and Statistical Aspects of Probability in Banach Spaces. Proceedings, 1985. Edited by X. Fernique, B. Heinkel, M.B. Marcus and P.A. Meyer. IV, 128 pages. 1986.

Vol. 1194: Complex Analysis and Algebraic Geometry. Proceedings, 1985. Edited by H. Grauert. VI, 235 pages. 1986.

Vol.1195: J.M Barbosa, A.G. Colares, Minimal Surfaces in $\mathbb{R}^3$. X, 124 pages. 1986.

Vol. 1196: E. Casas-Alvero, S. Xambó-Descamps, The Enumerative Theory of Conics after Halphen. IX, 130 pages. 1986.

Vol. 1197: Ring Theory. Proceedings, 1985. Edited by F.M.J. van Oystaeyen. V, 231 pages. 1986.

Vol. 1198: Séminaire d'Analyse, P. Lelong – P. Dolbeault – H. Skoda. Seminar 1983/84. X, 260 pages. 1986.

Vol. 1199: Analytic Theory of Continued Fractions II. Proceedings, 1985. Edited by W.J. Thron. VI, 299 pages. 1986.

Vol. 1200: V.D. Milman, G. Schechtman, Asymptotic Theory of Finite Dimensional Normed Spaces. With an Appendix by M. Gromov. VIII, 156 pages. 1986.

Vol. 1201: Curvature and Topology of Riemannian Manifolds. Proceedings, 1985. Edited by K. Shiohama, T. Sakai and T. Sunada. VII, 336 pages. 1986.

Vol. 1202: A. Dür, Möbius Functions, Incidence Algebras and Power Series Representations. XI, 134 pages. 1986.

Vol. 1203: Stochastic Processes and Their Applications. Proceedings, 1985. Edited by K. Itô and T. Hida. VI, 222 pages. 1986.

Vol. 1204: Séminaire de Probabilités XX, 1984/85. Proceedings. Edité par J. Azéma et M. Yor. V, 639 pages. 1986.

Vol. 1205: B.Z. Moroz, Analytic Arithmetic in Algebraic Number Fields. VII, 177 pages. 1986.

Vol. 1206: Probability and Analysis, Varenna (Como) 1985. Seminar. Edited by G. Letta and M. Pratelli. VIII, 280 pages. 1986.

Vol. 1207: P.H. Bérard, Spectral Geometry: Direct and Inverse Problems. With an Appendix by G. Besson. XIII, 272 pages. 1986.

Vol. 1208: S. Kaijser, J.W. Pelletier, Interpolation Functors and Duality. IV, 167 pages. 1986.

Vol. 1209: Differential Geometry, Peñíscola 1985. Proceedings. Edited by A.M. Naveira, A. Ferrández and F. Mascaró. VIII, 306 pages. 1986.

Vol. 1210: Probability Measures on Groups VIII. Proceedings, 1985. Edited by H. Heyer. X, 386 pages. 1986.

Vol. 1211: M.B. Sevryuk, Reversible Systems. V, 319 pages. 1986.

Vol. 1212: Stochastic Spatial Processes. Proceedings, 1984. Edited by P. Tautu. VIII, 311 pages. 1986.

Vol. 1213: L.G. Lewis, Jr., J.P. May, M. Steinberger, Equivariant Stable Homotopy Theory. IX, 538 pages. 1986.

Vol. 1214: Global Analysis – Studies and Applications II. Edited by Yu. G. Borisovich and Yu. E. Gliklikh. V, 275 pages. 1986.

Vol. 1215: Lectures in Probability and Statistics. Edited by G. del Pino and R. Rebolledo. V, 491 pages. 1986.

Vol. 1216: J. Kogan, Bifurcation of Extremals in Optimal Control. VIII, 106 pages. 1986.

Vol. 1217: Transformation Groups. Proceedings, 1985. Edited by S. Jackowski and K. Pawalowski. X, 396 pages. 1986.

Vol. 1218: Schrödinger Operators, Aarhus 1985. Seminar. Edited by E. Balslev. V, 222 pages. 1986.

Vol. 1219: R. Weissauer, Stabile Modulformen und Eisensteinreihen. III, 147 Seiten. 1986.

Vol. 1220: Séminaire d'Algèbre Paul Dubreil et Marie-Paule Malliavin. Proceedings, 1985. Edité par M.-P. Malliavin. IV, 200 pages. 1986.

Vol. 1221: Probability and Banach Spaces. Proceedings, 1985. Edited by J. Bastero and M. San Miguel. XI, 222 pages. 1986.

Vol. 1222: A. Katok, J.-M. Strelcyn, with the collaboration of F. Ledrappier and F. Przytycki, Invariant Manifolds, Entropy and Billiards; Smooth Maps with Singularities. VIII, 283 pages. 1986.

Vol. 1223: Differential Equations in Banach Spaces. Proceedings, 1985. Edited by A. Favini and E. Obrecht. VIII, 299 pages. 1986.

Vol. 1224: Nonlinear Diffusion Problems, Montecatini Terme 1985. Seminar. Edited by A. Fasano and M. Primicerio. VIII, 188 pages. 1986.

Vol. 1225: Inverse Problems, Montecatini Terme 1986. Seminar. Edited by G. Talenti. VIII, 204 pages. 1986.

Vol. 1226: A. Buium, Differential Function Fields and Moduli of Algebraic Varieties. IX, 146 pages. 1986.

Vol. 1227: H. Helson, The Spectral Theorem. VI, 104 pages. 1986.

Vol. 1228: Multigrid Methods II. Proceedings, 1985. Edited by W. Hackbusch and U. Trottenberg. VI, 336 pages. 1986.

Vol. 1229: O. Bratteli, Derivations, Dissipations and Group Actions on C*-algebras. IV, 277 pages. 1986.

Vol. 1230: Numerical Analysis. Proceedings, 1984. Edited by J.-P. Hennart. X, 234 pages. 1986.

Vol. 1231: E.-U. Gekeler, Drinfeld Modular Curves. XIV, 107 pages. 1986.

Vol. 1232: P.C. Schuur, Asymptotic Analysis of Soliton Problems. VIII, 180 pages. 1986.

Vol. 1233: Stability Problems for Stochastic Models. Proceedings, 1985. Edited by V.V. Kalashnikov, B. Penkov and V.M. Zolotarev. VI, 223 pages. 1986.

Vol. 1234: Combinatoire énumérative. Proceedings, 1985. Edité par G. Labelle et P. Leroux. XIV, 387 pages. 1986.

Vol. 1235: Séminaire de Théorie du Potentiel, Paris, No. 8. Directeurs: M. Brelot, G. Choquet et J. Deny. Rédacteurs: F. Hirsch et G. Mokobodzki. III, 209 pages. 1987.

Vol. 1236: Stochastic Partial Differential Equations and Applications. Proceedings, 1985. Edited by G. Da Prato and L. Tubaro. V, 257 pages. 1987.

Vol. 1237: Rational Approximation and its Applications in Mathematics and Physics. Proceedings, 1985. Edited by J. Gilewicz, M. Pindor and W. Siemaszko. XII, 350 pages. 1987.

Vol. 1238: M. Holz, K.-P. Podewski and K. Steffens, Injective Choice Functions. VI, 183 pages. 1987.

Vol. 1239: P. Vojta, Diophantine Approximations and Value Distribution Theory. X, 132 pages. 1987.

Vol. 1240: Number Theory, New York 1984–85. Seminar. Edited by D.V. Chudnovsky, G.V. Chudnovsky, H. Cohn and M.B. Nathanson. V, 324 pages. 1987.

Vol. 1241: L. Gårding, Singularities in Linear Wave Propagation. III, 125 pages. 1987.

Vol. 1242: Functional Analysis II, with Contributions by J. Hoffmann-Jørgensen et al. Edited by S. Kurepa, H. Kraljević and D. Butković. VII, 432 pages. 1987.

Vol. 1243: Non Commutative Harmonic Analysis and Lie Groups. Proceedings, 1985. Edited by J. Carmona, P. Delorme and M. Vergne. V, 309 pages. 1987.

Vol. 1244: W. Müller, Manifolds with Cusps of Rank One. XI, 158 pages. 1987.

Vol. 1245: S. Rallis, L-Functions and the Oscillator Representation. XVI, 239 pages. 1987.

Vol. 1246: Hodge Theory. Proceedings, 1985. Edited by E. Cattani, F. Guillén, A. Kaplan and F. Puerta. VII, 175 pages. 1987.

Vol. 1247: Séminaire de Probabilités XXI. Proceedings. Edité par J. Azéma, P.A. Meyer et M. Yor. IV, 579 pages. 1987.

Vol. 1248: Nonlinear Semigroups, Partial Differential Equations and Attractors. Proceedings, 1985. Edited by T.L. Gill and W.W. Zachary. IX, 185 pages. 1987.

Vol. 1249: I. van den Berg, Nonstandard Asymptotic Analysis. IX, 187 pages. 1987.

Vol. 1250: Stochastic Processes – Mathematics and Physics II. Proceedings 1985. Edited by S. Albeverio, Ph. Blanchard and L. Streit. VI, 359 pages. 1987.

Vol. 1251: Differential Geometric Methods in Mathematical Physics. Proceedings, 1985. Edited by P.L. García and A. Pérez-Rendón. VII, 300 pages. 1987.

Vol. 1252: T. Kaise, Représentations de Weil et GL_2 Algèbres de division et GL_n. VII, 203 pages. 1987.

Vol. 1253: J. Fischer, An Approach to the Selberg Trace Formula via the Selberg Zeta-Function. III, 184 pages. 1987.

Vol. 1254: S. Gelbart, I. Piatetski-Shapiro, S. Rallis. Explicit Constructions of Automorphic L-Functions. VI, 152 pages. 1987.

Vol. 1255: Differential Geometry and Differential Equations. Proceedings, 1985. Edited by C. Gu, M. Berger and R.L. Bryant. XII, 243 pages. 1987.

Vol. 1256: Pseudo-Differential Operators. Proceedings, 1986. Edited by H.O. Cordes, B. Gramsch and H. Widom. X, 479 pages. 1987.

Vol. 1257: X. Wang, On the C*-Algebras of Foliations in the Plane. V, 165 pages. 1987.

Vol. 1258: J. Weidmann, Spectral Theory of Ordinary Differential Operators. VI, 303 pages. 1987.

Vol. 1259: F. Cano Torres, Desingularization Strategies for Three-Dimensional Vector Fields. IX, 189 pages. 1987.

Vol. 1260: N.H. Pavel, Nonlinear Evolution Operators and Semigroups. VI, 285 pages. 1987.

Vol. 1261: H. Abels, Finite Presentability of S-Arithmetic Groups. Compact Presentability of Solvable Groups. VI, 178 pages. 1987.

Vol. 1262: E. Hlawka (Hrsg.), Zahlentheoretische Analysis II. Seminar 1984–86. V, 158 Seiten. 1987.

Vol. 1263: V.L. Hansen (Ed.), Differential Geometry. Proceedings 1985. XI, 288 pages. 1987.

Vol. 1264: Wu Wen-tsün, Rational Homotopy Type. VIII, 219 pages 1987.

Vol. 1265: W. Van Assche, Asymptotics for Orthogonal Polynomials VI, 201 pages. 1987.

Vol. 1266: F. Ghione, C. Peskine, E. Sernesi (Eds.), Space Curves Proceedings, 1985. VI, 272 pages. 1987.

Vol. 1267: J. Lindenstrauss, V.D. Milman (Eds.), Geometrical Aspect of Functional Analysis. Seminar. VII, 212 pages. 1987.

Vol. 1268: S.G. Krantz (Ed.), Complex Analysis. Seminar, 1986. VI 195 pages. 1987.

Vol. 1269: M. Shiota, Nash Manifolds. VI, 223 pages. 1987.

Vol. 1270: C. Carasso, P.-A. Raviart, D. Serre (Eds.), Nonlinea Hyperbolic Problems. Proceedings, 1986. XV, 341 pages. 1987.

Vol. 1271: A.M. Cohen, W.H. Hesselink, W.L.J. van der Kallen, J.F Strooker (Eds.), Algebraic Groups Utrecht 1986. Proceedings. XI 284 pages. 1987.

Vol. 1272: M.S. Livšic, L.L. Waksman, Commuting Nonselfadjoir Operators in Hilbert Space. III, 115 pages. 1987.

Vol. 1273: G.-M. Greuel, G. Trautmann (Eds.), Singularities, Repre sentation of Algebras, and Vector Bundles. Proceedings, 1985. XIV 383 pages. 1987.

Vol. 1274: N. C. Phillips, Equivariant K-Theory and Freeness of Grou Actions on C*-Algebras. VIII, 371 pages. 1987.

Vol. 1275: C.A. Berenstein (Ed.), Complex Analysis I. Proceedings 1985–86. XV, 331 pages. 1987.

Vol. 1276: C.A. Berenstein (Ed.), Complex Analysis II. Proceeding 1985–86. IX, 320 pages. 1987.

Vol. 1277: C.A. Berenstein (Ed.), Complex Analysis III. Proceedings 1985–86. X, 350 pages. 1987.

Vol. 1278: S.S. Koh (Ed.), Invariant Theory. Proceedings, 1985. V 102 pages. 1987.

Vol. 1279: D. Ieşan, Saint-Venant's Problem. VIII, 162 Seiten. 1987.

Vol. 1280: E. Neher, Jordan Triple Systems by the Grid Approach. X 193 pages. 1987.

Vol. 1281: O.H. Kegel, F. Menegazzo, G. Zacher (Eds.), Grou Theory. Proceedings, 1986. VII, 179 pages. 1987.

Vol. 1282: D.E. Handelman, Positive Polynomials, Convex Integr Polytopes, and a Random Walk Problem. XI, 136 pages. 1987.

Vol. 1283: S. Mardešić, J. Segal (Eds.), Geometric Topology ar Shape Theory. Proceedings, 1986. V, 261 pages. 1987.

Vol. 1284: B.H. Matzat, Konstruktive Galoistheorie. X, 286 page 1987.

Vol. 1285: I.W. Knowles, Y. Saitō (Eds.), Differential Equations an Mathematical Physics. Proceedings, 1986. XVI, 499 pages. 1987.

Vol. 1286: H.R. Miller, D.C. Ravenel (Eds.), Algebraic Topolog Proceedings, 1986. VII, 341 pages. 1987.

Vol. 1287: E.B. Saff (Ed.), Approximation Theory, Tampa. Procee ings, 1985–1986. V, 228 pages. 1987.

Vol. 1288: Yu. L. Rodin, Generalized Analytic Functions on Riemar Surfaces. V, 128 pages, 1987.

Vol. 1289: Yu. I. Manin (Ed.), K-Theory, Arithmetic and Geomet Seminar, 1984–1986. V, 399 pages. 1987.